ROUTLEDGE LIBRARY EDITIONS:
AGRICULTURE

Volume 13

# GENERAL AGRICULTURE FOR WEST AFRICA

# GENERAL AGRICULTURE FOR WEST AFRICA

Q. B. OLATUNJI ANTHONIO

Routledge
Taylor & Francis Group
LONDON AND NEW YORK

First published in 1979 by George Allen & Unwin Ltd

This edition first published in 2020
by Routledge
2 Park Square, Milton Park, Abingdon, Oxon OX14 4RN

and by Routledge
52 Vanderbilt Avenue, New York, NY 10017

*Routledge is an imprint of the Taylor & Francis Group, an informa business*



*British Library Cataloguing in Publication Data*
A catalogue record for this book is available from the British Library

ISBN: 978-0-367-24917-5 (Set)
ISBN: 978-0-429-32954-8 (Set) (ebk)
ISBN: 978-0-367-25002-7 (Volume 13) (hbk)
ISBN: 978-0-429-28554-7 (Volume 13) (ebk)

**Publisher's Note**
The publisher has gone to great lengths to ensure the quality of this reprint but points out that some imperfections in the original copies may be apparent.

**Disclaimer**
The publisher has made every effort to trace copyright holders and would welcome correspondence from those they have been unable to trace.

# General Agriculture for West Africa

Q. B. OLATUNJI ANTHONIO

London
GEORGE ALLEN & UNWIN
Boston Sydney

First published in 1979



GEORGE ALLEN & UNWIN LTD
40 Museum Street, London WC1A 1LU



**British Library Cataloguing in Publication Data**

**Anthonio, Quirino Bandele Olatunji**
**General agriculture for West Africa.**
**1. Agriculture**
**I. Title**
**630 S517.A/ 79-41076**

ISBN 0-04-631012-6

Typeset in 10 on 11 point Times
and printed in Great Britain
by Unwin Brothers Ltd, Old Woking, Surrey

*Contents*

## *Metric/Imperial Conversions*

| | | | |
|---|---|---|---|
| *Length* | | | |
| 1 cm | =0·39 in | 1 in | =2·54 cm |
| 1 m | =3·28 ft | 1 ft | =0·30 m |
| *Area* | | | |
| 1 $m^2$ | =10·76 $ft^2$ | 1 $ft^2$ | =0·09 $m^2$ |
| 1 ha | =2·47 acres | 1 acre | =0·40 ha |
| where 100ha | =1 $km^2$ | where 1 acre | =43,560 $ft^2$ |
| *Mass* | | | |
| 1 gm | =0·04 oz | 1 oz | =28·38 g |
| 1 kg | =2·20 lb | 1 lb | =0·45 kg |
| 1 tonne | =0·98 tons | 1 ton | =1·02 tonnes |
| where 1 tonne | =1,000 kg | where 1 ton | =2,240 lb |
| *Yield* | | | |
| 1 tonne/ha | =0·40 tons/acre | 1 ton/acre | =2·55 tonnes/ha |
| *Temperature* | | 5°F | =9°C+160 |
| 9°C | =5°F−160 | | |

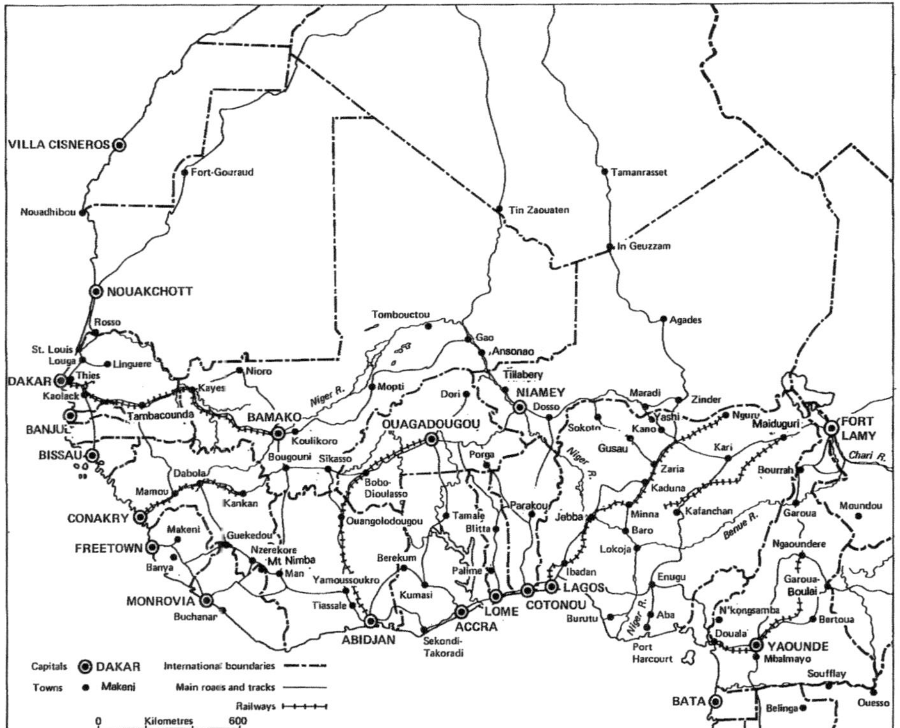


*Map of the West African sub-region*

*Part One*

# BACKGROUND TO TROPICAL AGRICULTURE

*Chapter 1*

# Climate, Soil, Man and Agriculture in West Africa

## 1.1 INTRODUCTION

It is always useful to start the study of any subject, particularly one as important as agriculture, with a definition. It must be understood, however, that it is very difficult to give a definition that will embrace the full meaning and scope of any term or subject. *Agriculture* may be simply defined as partly a science and partly an art, with both parts interacting in the cultivation of crops and the raising of livestock (and fish) either for direct consumption as food or for sale or for export. Crop cultivation, otherwise known as *crop husbandry*, involves the raising of field, garden and tree crops for food, animal fodder, manufacturing, processing or other purposes. Both annual crops (e.g. yams, beans, groundnuts, maize, vegetables) and perennial crops (e.g. fruit trees, cocoa, coffee, oil palm) are grown.

Agriculture also includes the raising of livestock (sometimes called *animal husbandry*) for milk (dairying); or for meat (beef rearing); or as work-bulls for ploughing or for other work and carrying loads (traction). Livestock such as chicken, guinea-fowl, ducks, cattle, sheep and goats are reared for eggs and meat. When a farmer raises livestock and at the same time cultivates crops for food or for sale the farmer is said to practise *mixed farming*, i.e. a combination of crop and livestock farming.

The role of man in agriculture is tied up with his basic need to survive. Man has basic needs for air, food, clothing and shelter, all of which can no longer be obtained free. Therefore, man over the ages has learnt the art of planting and harvesting crops and rearing livestock to supply his needs. Man has also scientifically raised seeds and has improved them to yield more and to keep them free from pests and diseases. The domestication of animals is a scientific effort to supply man with livestock to provide part of his food, shelter and clothing.

## 1.2 TYPES OF FARM PRODUCTS

To make the definition and scope of agriculture more meaningful, it may be helpful to say a few words about the nature and kind of products that are produced on farms. Farm products can be grouped for simplicity into two very broad categories. Products of direct plant origin are regarded as *crops* while animal products can be termed *livestock products.*

## 1.21 CROPS

There are different classes or groups of crops. Some are produced from plants that bear fruits and other products only once in a lifetime. These plants are called *annuals* if they are harvested completely within one calendar year (or season) of their being planted, or *biennials* if they live for two years. Examples of annuals are *field crops* such as yams, groundnuts, beans, maize and rice. Other annuals are *garden crops* such as okra, tomatoes, onions and other vegetables and ornamental plants (flowers such as roses, African marigolds, and so forth).

Other plants within the class of annuals and biennials are also cultivated, but not for food. The products from these plants are usually processed. Such plants are many and are of great economic importance in tropical Africa. They can be grouped into three broad categories: (a) *fibre crops* (e.g. cotton, jute plants such as kenaf); (b) *stimulants* (e.g. tobacco, spices such as pepper); and (c) *medicinal plants* (e.g. quinine tree).

There are some cultivated plants that take several years to reach the bearing stage and produce fruits for very many years thereafter. These farm crops are called *perennials* or permanent crops because, once they are at the bearing stage, they continue to produce fruits from one season to another for a good number of years – sometimes for as many as twenty to thirty years before they are no longer fruitful or die. There are four broad classes of perennials that are of importance to agriculturists: fruit trees, tree crops, fodder crops and forest trees.

Farms on which only *fruit trees* are cultivated are called *orchards.* The more common fruit in West Africa are mangoes, citrus (oranges), bananas and pawpaws. In most countries of West Africa fruit trees are more often cultivated as garden crops rather than planted on a large scale as orchards. On many farms fruit trees such as citrus, bananas and pawpaws are found planted along with other crops (mixed cropping of annuals and perennials).

Most *tree crops* – such as cocoa, the palms (coconut and oil palm), coffee and rubber – are also cultivated on small peasant farms as mixed crop or on plantations as sole crop. Tree crops are important because they produce fruits, seeds and other products some of which (e.g. oil palm fruits) can be sold locally while others (e.g. cocoa, coffee) are processed in industries locally or in other parts of the world, including countries like Britain, America, the USSR and the Netherlands where such crops cannot normally grow. For instance, the cocoa tree produces cocoa beans which are processed into chocolate, cocoa powder and cocoa liquor. These products of cocoa and other tree crops are in great demand all over the world.

*Fodder crops* are grown to feed livestock. The common saying that 'all flesh is grass' has a practical application in agriculture, because cattle, sheep and goats feed on grass to produce meat and milk for human consumption. Manure or excretory waste from cattle, sheep, goats, horses, donkeys and poultry is called farm yard manure (abbreviated FYM) and is extremely useful in soil regeneration to improve the fertility of the soil, thereby improving the yield of crops and grasses. Grass and some crops (tubers and cereals) are fed to livestock directly on the field as pasture. Sometimes the grasses are cut, brought in and fed to the animals in their stalls as fodder. Several tropical grasses are grown for fodder. The more common ones are guinea grass (*Panicum* spp.), elephant grass (*Pennisetum* spp.), Gamba or Sudan grass (*Andropogon* spp.) and giant star grass (*Cynodon* spp.).

The final group of perennials is *forest trees and shrubs*. The tropical world is very well known for its forest products, and African timbers are important and famous the world over. The more important ones include the iroko (*Chlorophora excelsa*), African mahogany (*Khaya* spp.), obeche (*Triplocohiton scleroxylon*), opepe (*Nanclea diderrichii*) and several others. Forests are very important, for many reasons including the following:

(1) Climatic effects: In particular, forests help to provide the environment necessary for cooling water vapour and inducing rain.

(2) Soil conservation: By checking the intensity of precipitation (rate of drop of rainfall on the ground), forests reduce the rate of flow of rain water, thereby reducing and sometimes (especially where there is thick undergrowth) eliminating erosion.

    The action of heat on soil is reduced because of the forest cover, and the fallen leaves provide water conservation and an ideal ecological environment for the rapid growth and development of soil micro-organisms, which produce humus.

(3) Forest products: These are very useful to man. Some important ones are:
    (a) logs for timber;
    (b) timbers for furniture and other constructions;
    (c) fibres for ropes, bags, etc.;
    (d) wildlife for meat, hides, skins and leather which can be made into bags or clothing materials or used for ornamental purposes;

(e) herbs, bark, leaves etc. that produce drugs for curing diseases or poisons for fishing, hunting or killing pests and wild animals;
(f) firewood for fuel for cooking and warmth;
(g) wood pulp for paper and other by-products such as dyes and resins which have various industrial uses.

(4) Wildlife conservation: This is useful for scientific information, for biological balance (preservation of natural ecology) and as a tourist attraction.
(5) Land reserve: Agricultural land reserve can be brought into cultivation if and when required.

### 1.22 LIVESTOCK

Livestock, or farm animals, refers mainly to four-legged or *quadruped* animals such as cattle, goats, sheep, horses and donkeys. Birds such as chickens, turkeys, ducks, geese and guinea fowls are also included. Fish is livestock but generally considered on its own as it is not part of a farm's regular stock of animals. However, some fishes are reared for their flesh, which is used for food, some for their oils, while others are used for medicinal purposes.

Four-legged farm animals can be divided into two broad groups. Those that feed mainly on vegetation constitute one group and are called *herbivores*, which are either *ruminants* (e.g. cattle, sheep, goats) or *non-ruminants* (e.g. horses, rabbits, donkeys). The other group is known as *non-herbivores* (e.g. dogs). Dogs, which are sometimes referred to as pets, can also be regarded as farm animals if they are used for hunting, as is common in Africa.

Livestock are very useful to man. They supply meat for human consumption (e.g. beef from cattle, mutton from sheep, game from poultry, pork from pigs and goat meat from goats) as well as milk (from cows). Poultry supply eggs, meat and feathers (the latter can be to stuff pillows, mattresses and cushions). Some animals (e.g. bullocks, horses, donkeys) are useful for work and for carrying farm produce and other loads. The coats of some animals (hides from cattle, skin from sheep and goats) are useful for leather making, while some animals (e.g. rabbits, dogs) are useful pets in West Africa.

## 1.3 CLIMATE

It is universally true, though often taken for granted, that climate and weather have a great impact on nearly everything that man does. His agriculture, clothing, style of housing and even hours of work, culture, laws and behaviour pattern are all affected by climate. The cycle of

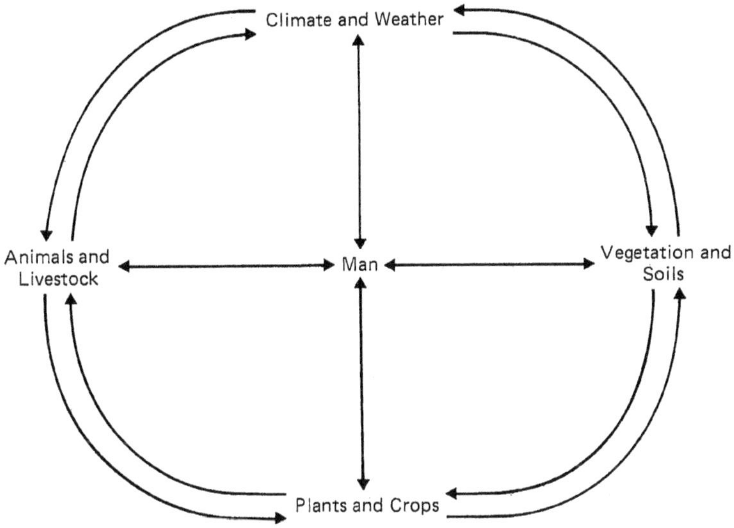


Figure 1.1 *The interdependence of nature and man.*

inter-relationship between man and his environment is illustrated in Figure 1.1.

The type and pattern of agriculture (farming and livestock raising) in any area is affected by a lot of factors which can be grouped into three broad categories: (a) natural factors, including soil, climate and other physical factors; (b) the technology (method) or the level of applied scientific knowledge and resources used in farming; and (c) social forces, including profitability and feeding habits. The second and third groups will be discussed later; here we start with climate.

### 1.31 FEATURES OF CLIMATE

In tropical Africa, as in any other part of the world, climate, soil and physical forces are very important in agriculture. It is probably easier to understand how these factors affect agriculture if we start by appreciating the facts that for their survival all animals require plant products as food, and that all plants in their turn require air, water and sunlight for growth. These three factors – air, water and sunlight – are supplied in any part of the country in varying amounts depending on the climate and the soil condition in which the plants grow.

The climatic conditions in tropical Africa can be discussed only briefly in this book and only in so far as these conditions affect agriculture. Any good geography book (e.g. *The Climate of Africa* by B. W. Thompson, OUP, 1965) will give a more explicit discussion of climate.

Climate includes temperature, rainfall, humidity, wind movements, length of daylight (or sunlight hours), evaporation and evapotranspiration. The two major climatic factors affecting agriculture are rainfall and temperature.

To start with, the tropical area is a very hot zone extending as a belt around the world on either side of the equator (or latitude 0°), to as far as the tropic of Cancer (latitude 23½°N) in the north and the tropic of Capricorn (latitude 23½°S) in the south. This belt in Africa passes through the Sahara Desert in the north, touching the northern boundary of the Republics of Niger and Chad, and through the Kalahari Desert in the south, cutting across Namibia (South West Africa), Botswanaland and the tail end of the Malagasy Republic. West Africa is wholly in the tropics, where it extends from 4°N to 18°N above the equator.

*Temperature* is one of the two most important climatic factors in West Africa. The equatorial region of West Africa is typified by a high temperature all year round, usually averaging 26–28°C (78–80°F). Because the sun is usually overhead in the tropics, there is little change in average temperature between day and night or between the dry and rainy seasons. However, there is usually a significant temperature difference of about 5–8°C (9–14·5°F) between sunlight and shade even in the same area. Temperature range becomes more pronounced moving away from the equator. In spite of this the temperature is very rarely above 37°C (100°F). It is never much below 21°C (70°F) either. Exceptions to this rule are found on the high peaks of the Cameroon mountains and on the Bauchic plateau in Nigeria where the temperature sometimes drops as low as 15°C (57°F). Table 1.1 gives the average monthly temperatures for some important cities of West Africa.

*Rainfall* is the other most important climatic factor, and two distinct types of rainfall pattern can be identified in West Africa: the *equatorial rains* with heavy downpour all year round, and the *tropical* (or *monsoon* type) *rains* where there is a distinct dry season.

In the equatorial region of West Africa the rains fall heavily at all seasons of the year. The rain is usually referred to as *convectional rain* whereby in the early part of the day the bright sunshine causes much evaporation. The ascending air is heavy with moisture which when cooled results in rain. Such rains are usually accompanied by thunder and lightning. The average annual rainfall can be as high as 200 cm (78 in).

In the tropical region of West Africa a distinct dry season begins to emerge. This season becomes more pronounced moving northwards

Table 1.1 *Average monthly temperatures in some West African cities, in °C (and °F).*

| *Month* | *Nigeria* | | | | *Ghana* | | *Ivory Coast* | | *Senegal* | |
|---|---|---|---|---|---|---|---|---|---|---|
| | *Lagos* | | *Kano* | | *Accra* | | *Abidjan* | | *Dakar* | |
| | Max. | Min. | Max. | Min. | Max. | Min. | Max. | Min. | Max. | Min. |
| January | 33 (91) | 20 (68) | 36 (97) | 9 (48) | 33 (91) | 23 (73) | 30 (86) | 23 (73) | 24 (75) | 18 (64) |
| February | 33 (91) | 22 (72) | 38 (100) | 11 (52) | 33 (91) | 23 (73) | 31 (88) | 24 (75) | 23 (73) | 18 (64) |
| March | 34 (93) | 22 (72) | 41 (106) | 14 (57) | 33 (91) | 23 (73) | 31 (88) | 24 (75) | 24 (75) | 18 (64) |
| April | 34 (93) | 22 (72) | 42 (108) | 19 (66) | 33 (91) | 23 (73) | 32 (90) | 25 (77) | 25 (77) | 19 (66) |
| May | 33 (91) | 22 (72) | 41 (106) | 21 (50) | 33 (91) | 23 (73) | 31 (88) | 24 (75) | 26 (79) | 20 (68) |
| June | 31 (88) | 22 (72) | 38 (100) | 19 (66) | 32 (90) | 22 (72) | 29 (84) | 23 (73) | 29 (84) | 23 (73) |
| July | 30 (86) | 22 (72) | 34 (93) | 19 (66) | 30 (86) | 21 (70) | 28 (82) | 23 (73) | 30 (86) | 25 (77) |
| August | 30 (86) | 21 (70) | 33 (91) | 18 (64) | 28 (82) | 21 (70) | 27 (81) | 21 (70) | 30 (86) | 25 (77) |
| September | 31 (88) | 22 (72) | 34 (93) | 19 (66) | 29 (84) | 22 (72) | 28 (82) | 22 (72) | 30 (86) | 24 (75) |
| October | 31 (88) | 21 (70) | 35 (95) | 16 (61) | 31 (88) | 22 (72) | 29 (84) | 23 (73) | 30 (86) | 25 (77) |
| November | 33 (91) | 22 (72) | 38 (100) | 13 (55) | 32 (90) | 23 (73) | 31 (88) | 24 (75) | 29 (84) | 23 (73) |
| December | 33 (91) | 21 (70) | 35 (95) | 11 (52) | 32 (90) | 23 (73) | 31 (88) | 24 (75) | 27 (81) | 20 (68) |
| Annual range | 4 (7) | 2 (4) | 9 (17) | 12 (22) | 5 (9) | 2 (3) | 5 (9) | 4 (7) | 7 (13) | 7 (13) |

away from the equator. Annual average rainfall is up to 180 cm (70 in) on the top edge of the equatorial region. This may then drop to 38 cm (15 in) on the edge of the desert.

The equatorial belt is characterised by the relatively wetter season. The early rains from approximately April to July mark the beginning of the planting season in the coastal areas of West Africa. The later rains or the second rainy season (also called 'minor rains') are from September to November. In the tropical climate belt there is only one rainy season, beginning usually from May nearer the coast and progressively later further inland. In Table 1.2 the average monthly rainfall figures are supplied for some important cities of West Africa. These confirm the discussion above.

Three other climatic factors – humidity, daylight hours and evapotranspiration – affect the vegetation, agricultural patterns as well as the crops and livestock in West Africa.

*Humidity* is defined as the degree of moistness or wetness of the atmosphere. A wet and dry bulb thermometer reading is used to determine humidity. In West Africa humidity is high, ranging from 85 to 95, and the atmosphere is said to be very damp.

Length of daylight, sometimes referred to as *sunlight periodicity* or *sunlight hours*, is an important climatic factor because it affects a number of physiological changes such as flowering and fruit setting in some farm crops. However, in West Africa, as in most tropical areas of the world, sunlight hours are usually within the same range throughout the year whereas they vary widely in other climates.

*Evaporation* and *evapotranspiration* signify the rate at which water is turned into vapour from one part of the earth's surface (or any object) and then released to form part of the vapour of the atmosphere. Evapotranspiration is the rate at which such change into vapour (*vaporisation*) is released during transpiration from a living organism. If a plant loses too much water it becomes weak and can no longer maintain an erect stance or shape and may fall. The plant under such a condition is said to be *wilted*, and it then collapses. It is always useful to supply water to a plant long before wilting sets in.

### 1.32 EFFECTS OF CLIMATE ON AGRICULTURE

The features of climate discussed above have very important effects on the soil as well as interacting with it to evolve different ecological zones in West Africa. In response to the combined effects of climate, soil and the ecology, man's need for food, shelter and income has prompted him to evolve the different types and patterns of agriculture – crop and livestock combinations – observed in different parts of West Africa.

Table 1.2 *Average rainfall in some West African cities, in cm (and inches).*

| Month | Nigeria | | | | Ghana | | Ivory Coast | | Senegal | |
|---|---|---|---|---|---|---|---|---|---|---|
| | Lagos | | Kano | | Accra | | Abidjan | | Dakar | |
| January | 0–5 | (0–2) | 0–5 | (0–2) | 0–5 | (0–2) | 0–5 | (0–2) | 0–5 | (0–2) |
| February | 0–5 | (0–2) | 0–5 | (0–2) | 0–5 | (0–2) | 0–5 | (0–2) | 0–5 | (0–2) |
| March | 10–20 | (4–8) | 0–5 | (0–2) | 5–10 | (2–4) | 10–20 | (4–8) | 0–5 | (0–2) |
| April | 10–20 | (4–8) | 0–5 | (0–2) | 5–10 | (2–4) | 10–20 | (4–8) | 0–5 | (0–2) |
| May | 30–40 | (12–16) | 5–10 | (2–4) | 10–20 | (4–8) | 30–40 | (12–16) | 0–5 | (0–2) |
| June | 40+ | (16+) | 10–20 | (4–8) | 10–20 | (4–8) | 40+ | (16+) | 20–30 | (8–12) |
| July | 30–40 | (12–16) | 20–30 | (8–12) | 0–5 | (0–2) | 20–30 | (8–12) | 20–30 | (8–12) |
| August | 5–10 | (2–4) | 30–40 | (12–16) | 0–5 | (0–2) | 0–5 | (0–2) | 30–40 | (12–16) |
| September | 10–20 | (4–8) | 10–20 | (4–8) | 0–5 | (0–2) | 5–10 | (2–4) | 20–30 | (8–12) |
| October | 20–30 | (8–12) | 0–5 | (0–2) | 5–10 | (2–4) | 20–30 | (8–12) | 10–20 | (4–8) |
| November | 5–10 | (2–4) | 0–5 | (0–2) | 0–5 | (0–2) | 10–20 | (4–8) | 0–5 | (0–2) |
| December | 5–10 | (2–4) | 0–5 | (0–2) | 0–5 | (0–2) | 10–20 | (4–8) | 0–5 | (0–2) |
| Annual total[a] | 165–250+ | (64–98+) | 75–155 | (29–60) | 35–105 | (14–41) | 155–245+ | (60–96+) | 120–185 | (47–72) |

[a] Discrepancies in total inches are due to rounding to the nearest whole number.

Table 1.3 *Climate, Vegetation and Agricultural Zones in West Africa.*

| *Vegetation Zone* | *Temperature* | *Annual Rainfall* | *Crops and Trees* | *Livestock* |
|---|---|---|---|---|
| *Evergreen equatorial* and *mangrove forest:* high humidity; evergreen forest; thick undergrowth. | Over 27 °C (80 °F); daily and annual range narrow, about ±5·5 °C (10 °F). | Over 254 cm (100 in); no distinct dry season. | Forest trees, rubber, oil palm, cocoa, coffee, banana. | Livestock not very important; goats mostly. |
| *Humid tropical:* mixed deciduous forest; forest more open. | 18–27 °C (65–80 °F); daily and annual range narrow, about ±5·5 °C (10 °F). | 203–254 cm (80–100 in); fairly distinct dry season; 2-peak rainfall distribution. | Mostly perennials, as above; annuals include tubers (e.g. yam, cassava), few cereals (e.g. maize). | Not much livestock; goats, pigs, poultry. |
| *Guinea savanna:* few trees; more grass; average high humidity. | 16–32 °C (60–90 °F); daily and annual range wide, about ±11 °C (20 °F). | 102–178 cm 40–70 in); dry season longer. | Wooded, savanna grasses, tubers (e.g. yam, cassava), cereals (e.g. rice, maize). | Sheep, goats, rabbits, poultry, cattle. |
| *Sudan savanna:* few or no trees, low humidity. | 10–32 °C (50–90 °F); daily and annual range wide, about 16·5 °C (30 °F). | 76–114 cm (30–45 in); prolonged dry season. | Cereals mostly (e.g. rice, maize, corn, millet). | Cattle, sheep, goats, poultry, horses, donkeys. |
| *Sahel savanna:* no trees; shrubs present. | 16–32 °C (60–90 °F); daily and annual range very wide, about ±19·5 °C (35 °F). | Less than 76 cm (30 in); short rainy season. | Short-season cereals and legumes mostly. | Cattle, sheep, poultry, horses, camels. |

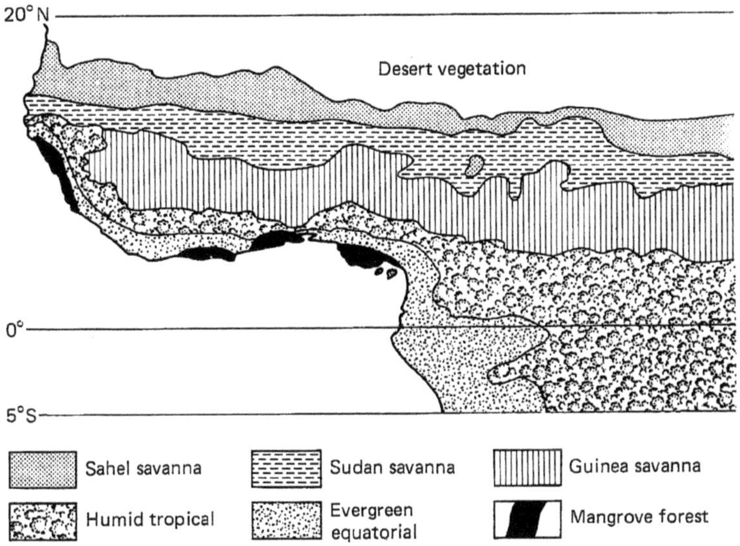


Figure 1.2 *Zonal distribution of ecology and agriculture in West Africa.*

Rainfall and temperature influence the amount of water available to plants. It is true that low precipitation can be counteracted by irrigation, the use of drought-resistant varieties, mulching and good husbandry to conserve moisture, but by far the most important item for plant growth is heat and moisture. Hence the ecological and agricultural pattern in West Africa is as indicated in Figure 1.2 and Table 1.3, where five agricultural zones are described and mapped. However, not all countries of West Africa have all five zones. Except for Nigeria, Dahomey and Togo, which have the five, most countries in West Africa have only two or three.

*Chapter 2*

# Land Use and Farming Systems in Tropical Africa

## 2.1 SYSTEMS OF LAND TENURE

Land use in any country is influenced by the social laws of the people. The rights, laws and social rules governing the use and/or ownership of land is called *land tenure*. Land tenure and climate influence land use especially in determining how much land is available to each farming household. The type of crops and livestock patterns are also influenced by these as much as by the food needs of the farmer's household.

There are different kinds of land tenure systems in West Africa, and in every case the tenure system affects the agricultural pattern. Only a few of the important ones are discussed here.

For very many years, land rights in West Africa have been communally owned. Under the system of *communal ownership* every adult citizen is entitled to a piece of land in his tribal area for the cultivation of his crops. In addition, everyone has the right to graze livestock on the communal pasture or grassland. Everyone has the privilege too of gathering wood, hunting and fishing in the communal forest or river (or stream). Communal ownership has the advantages of fair treatment and equal opportunity of giving land to every member of the society who wants to farm. There is some disadvantage in this system of land tenure in that some people who may not be using land may keep it idle and build a strong sentimental tradition of not wanting to give their right to someone else.

*Private individual ownership* is a distinctly different land tenure system by which the right to use land is recognised by law as being given to a specific individual instead of the community. It is argued that private land ownership has the advantage of encouraging the individual owner to be more productive because he reaps the benefits of his handiwork and of any improvements (or investments) that he may make. On the other hand, the disadvantages of individual ownership include land speculation, which may unnecessarily force up the price of land. Individual ownership may lead to a land shortage with increasing population and give rise to landless workers, thus creating social conflicts and other problems.

*Tenant ownership* is another form of 'ownership', where the land right is limited only to the right to use the land. In this case the ownership belongs to one person or group of persons (the *landlord*) who for one reason or another (money or other reward) allows another person (the *tenant*) the right to use the piece of land for a specific period. There are three important classes of tenancy:

(1) *Cash tenancy or leasehold*: The right to use the piece of land is given upon the payment of an agreed sum of money called the *rent*. The tenant pays the rent to the landlord. When payment is made with the products of the farm, it is called *payment in kind*. In some cases payment is made partly in cash and partly in kind; this is called *cash and kind tenancy*.
(2) *Land pledging*: The landlord has borrowed money (a *loan*) from someone and uses the land as a security for the loan. The landlord allows his benefactor (or *creditor*) the right to use the piece of land rent-free until the money lent to the landlord is repaid.
(3) *Share cropping*: The landlord is expected to help with some investment on the land, and the tenant pays back in kind by giving the landlord some agreed portion of the harvest. In some cases the tenant may, in addition, be expected to work or supply labour to the landlord's private farm. Very few of this type of tenancy are left in West Africa.

All types of land use rights have their advantages and disadvantages. However, they are all important because as a rule they influence the farming systems that obtain in any country. Furthermore, as the farming system changes, the system of tenure becomes modified as well.

It is therefore useful to study the different types of agricultural systems in West Africa.

## 2.2 AGRICULTURAL (FARMING) SYSTEMS

Agricultural systems are methods of organisations by which different groups of people employ the resources of land, climate, labour and capital or equipment to produce agricultural products such as crops or livestock or a combination of these. Agricultural systems often reflect the urge of human beings to produce both enough for their needs and also a surplus for sale, or perhaps to produce entirely for sale.

West African agriculture is largely adapted to these climatic, vegetation and tenancy characteristics. In the arid regions, for instance, low rainfall prevents the cultivation of crops on a large scale. Consequently the agricultural practice is *nomadic herding*, with the livestock roaming from place to place in search of grass and water. Where the vegetation can support livestock, *ranching* and/or *mixed farming* replaces nomadic herding.

*Shifting cultivation* is the agricultural system practised in many parts of Africa. A piece of land is cultivated for a few years, and when the soil is exhausted and crop yields are low the farmer abandons the farmland and then shifts his area of cultivation, sometimes building a small hut for shelter on the new plot. This system of farming is of course only possible where there is abundant land, a situation that is rapidly becoming untenable now in some parts of Africa because of increasing population. Shifting cultivation has some advantages. It checks the spread of crop pests and diseases and checks erosion as well.

*Bush fallow* is an integral part of shifting cultivation. A piece of farmland is left to rest (fallow), and very soon it acquires a regrowth of natural vegetation. When the vegetation is thick enough it may help to improve the fertility of the soil, thus it is useful as a means of soil regeneration.

Shifting cultivation, however, is the system of agriculture to be found in upland areas where leaching and erosion are a menace to soil, with bush fallow used to restore fertility. So long as land is abundant relative to population, the system achieves its primary object of restoring soil fertility. However, as the population increases, the period between successive cultivations is shortened, and consequently the system of checking erosion and restoring soil fertility is undermined.

In some parts of West Africa, the usually high fertility of the land and proximity to the market may compel peasants to stay put on the land and practise *extensive sedentary tillage*. Under this system the resting period of the land is shortened, and during such periods the peasants who are temporarily not farming while fertility is being restored, engage in other activities.

Another type of agricultural practice for which bush fallow is still used is *intensive sedentary tillage*. This is the system in highly fertile, though densely populated areas. In the sub-humid areas, given adequate supply of water during the growing period, intensive sedentary tillage has wide possibilities not only for crop production but also for animal husbandry, giving rise to mixed farming.

The most modern agricultural practice is found on *plantations* where a single crop is produced. This practice is called *mono cropping*, and farming is on a large scale to supply other markets. It is highly capitalised and requires mechanical and managerial skills sometimes imported from foreign sources. In comparison with peasant production, the plantation system has the advantages of a higher rate of technical improvements, greater financial resources and better marketing facilities. But its success depends on an expanding market, suitable technology, an adequate supply of relatively cheap labour, and a stable political environment. Unfortunately these variables tend to be very difficult to find all together in most countries of West Africa.

There are other types of modern agricultural systems. *Crop rotation*, for instance, is a system in which different crops are grown on the same piece of land in a logical or scientific sequence. Crop rotation has the advantages of maintaining soil fertility, reducing the incidence of pests and diseases as well as inhibiting excessive growth of weeds or parasites, while at the same time keeping the yield of crops at a reasonable level.

*Continuous cropping* is a modern system in which the same piece of land is cultivated year after year, in direct contrast to shifting cultivation in which the piece of land is temporarily abandoned after a few years of cultivation. Continuous cropping is only feasible with measures that ensure fertility maintenance, including the use of organic and inorganic manures, of fallows, and of farming methods such as rotational cropping. Under continuous cropping a good piece of farmland can maintain a relatively high density of population.

*Mixed farming* is a system of farming in which both crops and livestock are raised to complement one another. The inter-relationship in production makes it possible to cultivate pasture or maize to feed some of the animals being kept, and the animals in turn supply work (traction and load carrying) as well as manure. The manure (organic fertiliser) helps in improving soil conditions, and this increases the yield. The advantages of mixed farming include the following:

(1) Soil fertility is maintained. Farmyard manure (FYM) and other animal manure are supplied by sheep, goats, rabbits, poultry, cattle, etc. (including bone manure and blood waste manure). Also, with controlled grazing, animal deposits made during grazing improve the soil and shorten fallow periods.
(2) Animals act as scavengers: Feeding on farm wastes after harvesting or other by-products (e.g. bran, husks) and converting these to useful protein materials and other products.
(3) Labour is balanced all year round.
(4) Diet is improved, as animal products that are not marketed may be eaten domestically.
(5) Animals may supply power for ploughing, ridging and transport.
(6) The farm economy is rendered more stable. Income from the sale of animals or animal products, particularly in an emergency, minimises the uncertainties in farming caused by bad weather, pest attacks etc.
(7) Higher productivity is achieved per unit of input (land or labour). Experiments in West Africa and Uganda, East Africa, have shown

that a grazed ley is effective in building up soil fertility, with increased yield of subsequent crops.

*Mixed cropping* is an agricultural system in which several crops are cultivated simultaneously on the same piece of land. Usually there is a major and a minor crop. The advantages of mixed cropping include security of food and of income and the maintenance of fertility. However, yields are usually lower for each crop. Among the advantages are:

(1) The shallow and deep-rooting plants in a mixture tap different zones in the soil, hence making efficient use of land for the season.
(2) One crop may help to prop another.
(3) A cereal benefits by deriving nitrogen fixed by a legume growing along with it.
(4) An unforeseen wet season may suit one crop and a dry season suit the other crop; thus total failure in a bad season can be avoided.
(5) Insects or disease may attack one crop and leave the other unscathed.
(6) In the case of mixed farming, animals usually thrive better on mixed fodder than on single plant feed.

*Part Two*

# THE PRODUCTION AND MANAGEMENT OF TROPICAL CROPS

*Chapter 3*

# The Botany and Management of Tropical Crops

## 3.1 INTRODUCTORY BOTANY

### 3.11 MAIN GROUPS OF CROP PLANTS

Generally, all living things are grouped into two broad categories or *kingdoms* known as the plant and animal kingdoms. All plants are divided into different *families*, and every family is further divided into *genera*, and within each genus are several *species*. Within each species are different *kinds* and *types*. The bases of most differentiation between plants and of their classification are their external and internal features, their flowers and their flowering habits.

In agriculture, the farmer is more concerned with only a limited number of plants that he can cultivate and use either for food or for other purposes. These are called *crops*. The more important tropical crop plants of West Africa can be grouped into the following categories based on their botany, food value and economic use and the nature of the product.

*Cereals* (Family Graminae)
The most widely cultivated cereals in Africa include maize (*Zea mays* Linn.), millet (*Pennisetum* sp.), guinea corn (*Sorghum* sp.), rice (*Oryza sativa* Linn.), wheat (*Triticum* sp.), acha or hungry rice (*Digitara* sp.) and finger millet (*Eleusine coracana* Gaertin).

Cereals are very important food grains because of the high proportion of concentrated starch contained in their seeds. Cereals also contain varying amounts of protein and minerals more than any other group of starchy products. They are used as human and livestock feed.

*Legumes* (Family Leguminoseae)
These are also sometimes called *pulses*. All legumes are useful sources of vegetable proteins for human as well as animal consumption. Legumes are also very important in agriculture because of the nitrogen fixation ability in their root nodules through the action of *Rhizobium* sp. bacteria.

The more important legumes of West Africa are cowpeas or black-eyed beans (*Vigna unguiculata* (L.) Walp), groundnuts (*Arachis hypogaea* L.), soya beans (*Glycine maximus* Merr), green gram (*Phaseolus aurens* Linn.), bambara groundnuts (*Voandzeia subterranea* Thouars), lima beans (*Phaseolus lunatus* L.) and pigeon peas (*Cajanus cajan* Druce).

*Roots and tubers*
Some tropical plants store part of their carbohydrate products in their roots or underground stems (called *tubers*). Some of these stored starch plants have proved edible and have for centuries become the main source of calories (starchy foods) for man and his domesticated animals.

The more important tuber crops in West Africa are yams (*Dioscorea* sp.), cocoyams (*Xanthosoma sagittifolium* (L.) Schott and *Colocasia esculenta* Schott). The more important root crops are cassava (*Manihot utilissima* Pohl), sweet potatoes (*Ipomea batatas* Poir) and tigernuts (*Cyperus esculentus* L.).

*Vegetables*
These constitute another major crop classification. Vegetables are leafy crops which in West Africa are usually cooked with other crops, animal products and spices into 'stew' and then eaten with the starchy portion of the local diets.

The more important tropical vegetables are onions, shallots, cucumbers, tomatoes, amaranthus, okra (*Hibiscus esculentus*), talinum (Yoruba – *gburẽ*), bitter leaf, telfaria, vegetable jute (*Corchorus* sp.) and egg plants. In addition, a few vegetables of the sub-tropical and temperate areas have been successfully cultivated in West Africa. The more common ones include carrots, spinach, lettuce, cauliflowers, radishes and cabbage.

*Edible fruits and nuts*
The hot and humid climate of West Africa is most suitable for a wide variety of tropical fruits and nuts. Unfortunately most of these fruits are not cultivated on a commercial basis in very many parts of West Africa. The more common and important ones include citrus fruits (*Citrus* spp.), bananas (*Musa sapientum* L.), plantains (*Carica paradisiaca* L.), pawpaws (*Carica papaya* L.), pineapples (*Ananas comosus* L.), cashew nuts (*Anacardium occidentale* L.), mangoes (*Magnifera indica* L.), avocado pears (*Persea americana* Mill) and guavas (*Psidium guajava* L.).

In most parts of West Africa, these fruits and nuts are used for food and to supplement the supply of mainly oils, sugars, minerals and vitamins in diets. Most of them (e.g. citrus, bananas, mangoes) are eaten in their natural state. Some are canned (e.g. mangoes, pineapples) and sold as such. Others (e.g. citrus, pineapples) have their juice extracted and bottled or canned (fruit juice). Still others (e.g. plantains) are boiled, fried or roasted for food or prepared in other forms.

*Beverages, spices and drugs*

For as far back as man can remember, West African forests and farms have provided man all over the world with a wide variety of beverages (cocoa, coffee, kola), spices (chillies, pepper, bean seed, ginger, sugar cane) and drugs which can be extracted from parts of plants, such as tobacco leaves (nicotine), kola fruits, coffee beans (cafeine) and tree bark (quinine).

The beverages cocoa (*Theobroma cacao*) and coffee (*Coffee robusta* and *C. arabica*) are cultivated in the wetter (humid tropical) parts of West Africa. The cocoa pods and coffee berries are harvested when ripe, and the beans are cleaned, fermented, dried and roasted before extracting the cocoa and coffee powder which are used as beverages.

Spices such as chillies, onions (*Allium cepa* L.), shallots (*Allium ascalonicum* L.), peppers (*Capsicum* sp. L.), ginger etc. are used in either their fresh or their dried state (e.g. ginger, sugar cane) or cooked into meat, fish and other food preparations (e.g. pepper, chillies, ginger). Some of the drugs, e.g. nicotine from tobacco (*Nicotiana tabacum*), are only taken into the body after processing (e.g. from cigarette, cigar and pipe tobacco).

The tremendous financial contribution made to the foreign exchange earnings of some West African countries – Ghana, Ivory Coast and Nigeria, to name a few – by the export of these agricultural products explains why so much effort is, and why more should be, devoted to their improvement, cultivation, harvesting, disease control and export promotion. Tobacco, for instance, is an important agricultural product in many countries of West Africa, because the processing of the leaves after harvesting into cigarettes and cigars provides local employment which helps economic development just as much as the income received by the farmers who cultivate the tobacco plants.

*Oils, latex and fibres*

Another group of important agricultural commodities exported by a number of West African countries includes oils, latex and fibres. The more important oil seeds come from the palm tree (*Elaeis guineensis* Jacg.) producing palm oil and palm kernel oil, groundnuts (*Arachis hypogea* L.), the coconut palm (*Cocos nucifera* L.) producing coconut oil from copra, and the shea-butter tree (*Butyrospermum parkii* Kotschy). The shea-butter is the principal source of edible oil in the drier savanna parts of West Africa where the oil is also used for fuel and lighting, in lamps. In addition to exporting the oil seeds themselves, the oils extracted are used locally in different food preparations and in cosmetics and pommades.

Rubber (*Hevea brasiliensis* (Kunth) Müll–Aug.) is the main latex-producing crop exported from West Africa. It is usually cultivated on plantations and peasant farms in the wetter tropical areas of West Africa. It is also processed locally and used extensively in all its types and forms, thus providing factory employment and substantial income for the peasant producers.

Some fibre crops are equally important export crops. An important one is cotton (*Gossypium* spp.). Kapok from the silk cotton tree (*Ceiba pentandra* L.) is the floss from the fruit and is a useful fine cotton. Other important fibres are raphia (*Raphia hookeri* Mann & Wendl), jute (*Urena lobata* L. or *Corchorus olitorius* L.) and sisal hemp (*Agave sisalana* Perrine), most of which do not enter into any elaborate commercialisation, even though the fibres are used extensively in bags, baskets and other containers as well as in local crafts and other manufactured goods.

## 3.12 USES OF CROP PLANTS

(1) *Food:* The most basic use of crop plants is for human food. It may be eaten either in its raw state, like many fruits (e.g. mangoes, citrus, pineapples), or when boiled or cooked (e.g. rice, yams, cassava). Sometimes processing is called for as a means of preservation, producing forms such as cassava grains (or gari), maize flour, bean flour, cassava and yam flour etc. These preparations can be made into pastes, cakes, puddings etc. when required. Several types of vegetables are cooked with special condiments to make different kinds of soups and stews.

(2) *Beverages and stimulants:* Cocoa, coffee beans and other seeds are fermented, roasted and processed to make the brews that take these names. With tea, not grown in West Africa, it is the tea leaves that are brewed into a stimulant. Other fruits and spices (e.g. ginger, kola) are eaten in their raw or partly cured state and consumed as stimulants.

(3) *Clothing:* cotton and other fibres are used in the making of clothing, tarpaulins and storage articles such as bags and satchels.

(4) *Industrial use:* Crops such as tobacco leaves, tobacco seeds, rubber latex, fibres, oil seeds etc. are of great industrial use both locally and overseas.

(5) *Fuel, light and energy:* Oil seeds produce oils of different types which can be used for light. Other plants can be used as fuel (e.g.

oil fibre, firewood). Wood is used in the form of charcoal (carbonated wood) to supply energy.

(6) *Drugs:* The successful development of the traditional herbalist in West Africa is mainly due to the profound use made of fruits, leaves, stems and the roots of different plants in their raw, boiled or charred state.

(7) *Animal feed:* Crops used for animal feed are mostly vegetative, either in the form of grasses (pasture) or as concentrates from the fruits and seeds of plants such as cotton, oil palm (kernel) and groundnuts. Some plant roots (e.g. cassava, potatoes) are also useful animal feed. Other crops (e.g. maize, molasses) are made into silage or hay for stock feed.

(8) *Organic manure:* Composts are essentially plant residues that have decomposed. Green manures are legume plants (e.g. beans, mucia, calapaganium) that are buried to supply nutrients, mainly nitrogen, to the soil.

(9) *Hedges and fences:* Occasionally live plants are used as hedges, while the stems of other plants (e.g. teak) provide good poles for fences, electric poles etc.

(10) *Constructions:* In buildings and other construction work, plant stems (e.g. coconut) as well as forest products (timber) are useful building materials. Other woods are used in furniture.

(11) *Ornaments:* Many flowers and other shrubs are useful as ornaments because of either their decorative appeal (colour, shape) or their aromatic scent. Others are used as extra scents for perfumes. Still others make good ornamental hedges.

(12) *Other uses:* Cosmetics, toiletries, dyes, poisons (for fish catching) etc. are among a host of other minor uses which cannot be listed here.

## 3.13 PLANT STRUCTURE

The crop plant as farmers know it is a living organism of the vegetable kingdom as distinct from animals belonging to the animal kingdom. The special characteristics of plants are: (a) they are capable of living wholly or partly on inorganic substances; and they have neither (b) power of locomotion (i.e. cannot move from place to place); nor (c) special organs for sensation; or (d) for digestion.

All plants have at least three parts:

(1) *The root:* This is usually underground even though a few plants (e.g. some yam varieties) have aerial roots (in the air or above ground).

(2) *The stem:* This is usually above ground level even though a few plants (e.g. grasses, yams) have underground stems.
(3) *The leaves:* These are always above ground and usually, though not always, green in colour.

*The Root*

Nearly all plants have different types and sizes of roots. The different types of roots on any one plant are collectively called the *root system*.

The functions of the root system are (a) to hold the plant erect in the ground (i.e. anchor it); (b) to absorb water and soil nutrients for plant use; and (c) in some plants to hold a store of food (e.g. cassava). The most important foods in reserve are usually starches (carbohydrates), which are used for feeding both animals and man.

There are five main groups of root system: *fibrous* roots (e.g. cereals such as rice and maize; grasses), *creeping* roots (e.g. citrus), *tuberous* roots (e.g. cassava), *tap* roots (e.g. cocoa, coffee) and *adventitious* roots (e.g. oil palm, rice, maize, mangrove, bamboo).

There are three main parts of the mature root, which can be seen with a microscope when a section is cut along a horizontal plane (a *cross-section*): The outer layer or *epidermis*; the next layer of tissue called the *cortex*; and the inner layer of tissue called the *vascular tissue*, which consists of the *phloem* and *xylem vessels*. (Readers should revise their botany for the functions of these important tissues.)

It is important that, for a crop plant to grow well, it must have a well-developed root system and good supply of nutrients from the soil (i.e. the soil has to be rich).

*The Stem*

The stem has several functions: (a) to bear the leaves, flowers and fruits; (b) to serve as a channel for conducting to the leaves the water and mineral salts which the roots absorb from the soil and for conducting the food manufactured in the leaves to other parts of the plants for their use; and (c) in some crop plants (e.g. yams, sugar cane), to store food.

Plant stems exhibit different external forms, which can be described as: *trailing* on the ground (e.g. groundnuts; most legume crops including centrosema, marrows, water melons and cucumbers), or *climbing* (e.g. yams, cowpeas), or *upright* (e.g. cassava, cocoa, coffee, rubber, cotton; cereals such as maize, rice, millet and guinea corn), or *partly underground* (e.g. onions and sweet potatoes).

When a cross-section is viewed under a microscope, the stem is seen to be composed of four main layers of tissues: the outer layer or *epidermis*, which in some old plants is sometimes protected by an outer layer of dead cells which together with the epidermis is referred to as *bark*; an inner layer of *cortex*; the central *pith*; and the *vascular bundles*, which are specialised cell groups arranged between the cortex and the pith.

*The Leaf*

One of the most important functions of the leaf is the manufacturing of carbohydrates using the sunlight and food nutrients (or *sap*) absorbed by the roots. Leaves of most crop plants are green in colour and are attached to the stem by the leaf stalk, called a *petiole*. Some leaves of grasses (and cereals) do not have petioles. Such plants have *leaf sheaths* which are wrapped round the stem, but whether with petiole or sheath the leaf is connected to the stem. The petiole extends far into the leaf as the *midrib*. The midrib in turn is divided into *veins* – the primary and secondary veins. The petiole or sheath, midrib and veins serve as the transportation network of the leaves. Between the leaf veins the leaf is flat and relatively darker; this portion of the leaf is known as the *lamina*.

The leaves are very useful for identification and classification purposes. Useful features are:

(1) *Size:* Leaves may be large or small.
(2) *Petiole or sheath:* Some plants (e.g. guava) have short petioles; others (e.g. pawpaw) have very long petioles.
(3) *Number and arrangement of veins:* Veins may be arranged in parallel (e.g. rice, maize, grasses) or as a network branching from the midrib (e.g. cotton, yam).
(4) *Position of leaves on the stem:* The arrangement of the leaves may be alternate (e.g. yam), opposite (e.g. coffee) or *verticillate* (e.g. allamerda or almond). Verticillate arrangement is when three or more leaves appear at the stem height point of the plant.
(5) *Leaf form:* This may be *simple*, i.e. with only one leaf at the point of the stem (e.g. rice, millet, yam), *lobed simple* (e.g. cassava, cotton) or *compound* (e.g. beans, cassava) with several small 'leaves' called *leaflets* growing out of one petiole.
(6) *Leaf edge:* This may be smooth (e.g. orange) or serrated, i.e. like a saw edge (e.g. okra).
(7) *Leaf surface:* This may be hairy (e.g. tobacco), rough (e.g. sand paper leaf: Yoruba – *epin*) or smooth and glossy (e.g. orange).
(8) *Thickness:* Leaves may be thin and sometimes transparent (e.g. tomato) or thick (e.g. mango).

A cross-section through the *leaf lamina* viewed through a lens or microscope will show an outer layer (top and bottom) of *epidermal* cells, closely followed from the top by a layer of *pallisade tissue* and then by the *spongy tissue* which contain air spaces. A cross-section through the midrib and veins will also show the *xylem* and *phloem* cells.

The function of food synthesis in the leaf is carried out in the green *chloroplasts*. These contain *chlorophyll*, the green pigment without which no carbohydrates synthesis can be performed. Most of these chloroplasts are in the pallisade tissue of the leaf lamina.

## 3.14 REPRODUCTION

### *The Flower*

All living things reproduce. With plants the reproduction is either sexual or asexual. The sexual organ of reproduction in plants is the *flower*. Flowers are produced as *buds* from the plant stem. Some plants produce the male flowers separately from the female flowers; in some the two types of flowers are on separate parts of the plant (e.g. maize, millet), while in others the two types of flowers are on two different plants (e.g. pawpaw). Most plants, however, have both the male and female organs on the same flower (e.g. okra).

The sexual parts of the flower are the *androecium*, made up of the *stamens* or male organs, and the *gynoecium*, made of the *carpels* or female organs. The other parts of the flower are accessories. Some parts (e.g. the *calyx* and *sepals*) protect the plant; others (e.g. the *petals*) are for attracting birds or insects which help with pollination.

A stamen is composed of a tubular *filament* and of *anthers* in which pollen containing the male gametes is produced. The female organ or carpel is more complicated that the male organ. It consists of an *ovary* in which the *ovules* (female eggs) are developed and stored until fertilisation, a tubular *style* and a *stigma* (the pollen receptacle).

When pollen is transferred to a stigma by either a bird, an insect or the wind, this is called *pollination*. Successful pollination leads to the process of *fusion* in which the pollen tube grows through the style of the female organ right down to the ovary and then to the ovule. The nucleus of the male gamete fuses with the nucleus of the female gamete to form a *zygote*. This fusion of the male and female nuclei is known as *fertilisation*.

The fertilised ovules develop into *seeds*, and the ovary after fertilisation develops into the *fruit*. If the ovary has only one ovule the fruit will have only one seed (e.g. mangoes, avocado pears). However, if the ovary has several ovules the fruit will have several seeds (e.g. guavas, okra, beans, maize). Some crop plants that are of great commercial value (e.g. oranges, tomatoes) can be bred so as to produce no seeds (aborted) even though the fruits are well formed. And in some cases the male flowers are sterile in the sense that they cannot fertilise the ovules (e.g. some flower plants).

*Fruits and Seeds*

Fruits and seeds develop from and after the process of fertilisation. A seed has several parts. These can be easily identified when the seed is cut along its *vertical axis*, which is defined as the straight plane from the point of seed attachment to the top of the seed. There is an outer protective coat known as the *testa*. There are the *cotyledons*, which are the first leaves of the plant; some crop plants (e.g. beans) store food in the cotyledons for use when the seed starts to germinate. It is important to note that, although food for the young plant is sometimes stored in the *endosperm* (e.g. maize and other cereals), the cotyledon functions in transferring food to the growing plant. Some seeds (e.g. grasses; cereals such as maize, rice, millet and guinea corn) have only one cotyledon; such plants are said to be *monocotyledonous*. Others with two cotyledons (e.g. beans, coffee, cocoa) are called *dicotyledonous*. Seeds also have a *plumule*, which upon *germination* of the seed forms the first plant foliage leaves, and a *radicle* which later develops to form the root system of the plant.

*Germination*

For seeds to germinate, four essential conditions must be fulfilled: (a) the seed must be viable, i.e. alive and in good condition; and there must be (b) moisture; (c) warmth; and (d) sunlight.

Farmers usually plant seeds after the first few showers of rain, provided the soil is wet enough. Seeds are planted at the appropriate depth and well covered to give warmth as well as protection from pests such as rats, birds and insects. If previous planting in a nursery is required, then the seed bed has to be specially prepared, and water is usually added immediately after planting.

The first part of the seed to emerge is the radicle, which points downwards (a condition described as *positively geotropic*), burying itself further into the soil. Soon after, *root hairs* grow out of the radicle, and the process of nutrient absorption starts. Since nutrients are absorbed only in liquid form, the presence of water for germination is vitally necessary.

The next part of the seed to emerge is the plumule, which points upwards against gravity (*negatively geotropic*) and above the ground towards light (*positively phototropic*). In some plants (e.g. beans and other legumes) the plumule lies between the cotyledons, and they invariable emerge together. In other plants, especially cereals (e.g. maize, millet), the plumule is protected by a sheath called the *coleoptile*, which soon drops off or unfolds as the leaves emerge.

## 3.15 PLANT GROWTH AND NUTRITION

Growth is the logical sequence to germination. With growth also comes the development of the plant. It is in the good growth and development of crop plants that farmers' interests and rewards lie. In other words, for a good harvest the farmer needs a good yield of crops, which is only possible if there is sufficient growth and adequate development of the plants.

For good plant growth and development, the four essential forces at work are:

(1) *The weather:*
   (a) sufficient and timely rain;
   (b) enough sunlight or periodicity of sunlight;
   (c) heat moderately adequate – scorching heat leads to a high rate of evapotranspiration and plant wilting, but enough heat is required to dry most grain crops and for plant survival.

(2) *The soil:*
   (a) good soil structure and the presence of adequate water and of soil-activating micro-organisms;
   (b) good soil texture and its ability to sustain and hold the plant erect and to allow adequate levels of air and water penetration;
   (c) soil nutrients (soil fertility) present in optimum amounts. Plants require some nutrients in large amounts (e.g. nitrogen, N; phosphorus, P; potassium, K; calcium, Ca; magnesium, Mg; sulphur, S). These are sometimes referred to as *macro nutrients*. The *micro nutrients* (e.g. iron, Fe; manganese, Mn; boron, B; copper, Cu; zinc, Zn; molybdenum, Mo) are just as vital but they are required only in relatively smaller quantities. Other minerals (e.g. sodium, Na; silicon, Si; chlorine, Cl) are useful to some particular plants.

(3) *Field husbandry:*
   (a) use of high-yielding strains of crop seeds and livestock;
   (b) correct seasonal timing of farm operations;
   (c) correctly carrying out all routine cultural operations (e.g. planting, transplanting, application of fertiliser, weeding, topping, staking, harvesting, storage);
   (d) use of correct implements and methods in farm operations;
   (e) maintaining good farming husbandry generally (e.g. crop rotation, correct seeds, correct dosage of chemicals for the control of diseases).

(4) *Farm management:*
   (a) adequate investment for the supply of necessary inputs at appropriate time in desirable quantities;

(b) effective planning and co-ordination of all farm operations;
(c) keeping records of physical factors as well as of financial revenue and expenditure on the farm – this helps to identify the weaker areas of the farm for subsequent improvement, especially in the light of changing market and economic conditions.

The process of growth and development of the crop plant is identified with a number of activities within the plant itself. A thorough knowledge of these activities helps to understand and improve such activities to increase crop yields. The major activities include: (a) plant nutrition and the nitrogen cycle; (b) photosynthesis and the carbon cycle; (c) plant food distribution and storage; (d) respiration and transpiration; and (e) farm cultural practices. These five activities will now be discussed in more detail.

### *Plant Nutrition and the Nitrogen Cycle*

An analysis of the component parts of the plant would probably reveal that plants are composed of water and dry matter in the following proportions:

water: 80–85 per cent of plant weight
dry matter: 15–20 per cent of plant weight

The dry matter is made up mainly of carbon, C, oxygen, O, and hydrogen, H. Minerals constitute about 1·5–2 per cent of the plant weight. Table 3.1 gives a brief summary of the functions of the major nutrients, as well as the deficiency symptoms that may arise when the nutrient is in short supply or completely not available. For most soils in West Africa continuous cropping reduces the presence and availability of nutrients. Deficiency symptoms can be corrected by the use of organic manure (e.g. copra, FYM or burned legumes) or fertilisers.

One very important element which is readily lost on farmland subject to continuous cultivation is *nitrogen*. Nitrogen in the soil is lost through leeching, exposure to air, and plants using it in their body cells and tissues. The phenomenon by which a recycling of this element is brought about in nature is useful indeed and must be discussed.

Most of the activities in the *nitrogen cycle*, shown in Figure 3.1, involve the conversion of atmospheric nitrogen into *nitrates* mainly by *nitrogen-fixing bacteria*, most of which live a symbiotic life in the roots of leguminous plants (e.g. cover crops such as centrosema; groundnuts, beans, peas). The nitrates so produced are subsequently absorbed by plant roots and used to build up *protein* compounds in plant tissues.

water absorbed by the roots, in combination with the carbon dioxide of the atmosphere. This process is known as *photosynthesis*. Air, and in particular carbon dioxide, enters the leaves through the openings (*stomata*) between the *guard cells* of the leaves. The chlorophyll in the chloroplasts is able to absorb the energy of direct sunlight and use it for photosynthesis.

Photosynthesis is very important and complicated. The first compound formed is usually a type of simple *sugar* which is later converted into *starch*. Oxygen is a by-product of this synthesis which is in form of the following equation:

$$6CO_2 + 6H_2O \xrightarrow[+\text{chlorophyll}]{\text{light energy}} C_6H_{12}O_6 + 6O_2$$

Once the starch is formed, depending on the plant needs the starch may be stored in specific storage organs (e.g. yam tuber, cassava roots). Alternatively, the starch through *enzyme* action may be converted into sugar which is then transported to any and all parts of the plant needing it.

The *carbon cycle*, the process by which carbon dioxide is recycled in nature, is illustrated in Figure 3.2. It can be seen that *respiration*

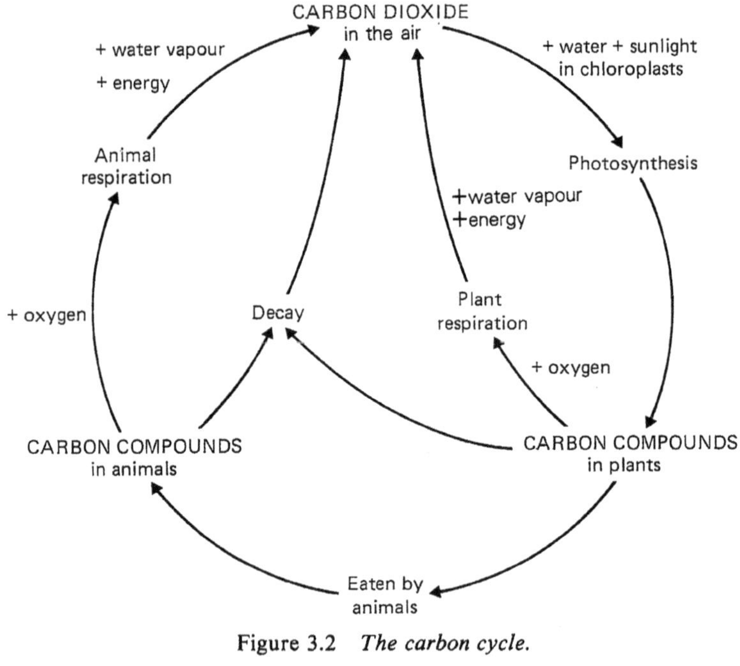


Figure 3.2 *The carbon cycle.*

'undoes' the work of photosynthesis, by using oxygen to 'burn up' carbohydrates and so release the carbon dioxide, water and energy originally taken in by the plants.

*Plant Food Distribution and Storage*

Carbohydrates are the most common food produced by all green plants. Through enzyme action, the starch is processed into sugar and distributed throughout the plants as required. When more is produced than is immediately required by the plants a method of storing the surplus has to be evolved. The food product stored varies between plants.

Plants have different *storage organs*, some of which are consumed by animals and humans before the plants can use them later. The major storage organs of crop plants in West Africa include the following:

(1) *Tubers:* The underground stem of the plant is converted into a storage organ (e.g. yams). Food is usually stored in the form of starch.
(2) *Roots:* The roots are converted into storage organs. Food is stored either as starch (e.g. cassava) or as sugars (e.g. sweet potatoes).
(3) *Leaves:* Some plants (e.g. onions, tobacco) keep their reserve in the leaves. The stored products vary between different plants; e.g. onions store sugars, other carbohydrates and etheral oils, whereas tobacco stores nicotine.
(4) *Stems:* In the case of sugar cane it is sugar that is stored in the stem. Cactus stems tend to be flat and juicy because they are filled with stored food of different combinations, some of which may be poisonous. In rubber stems, latex is stored.
(5) *Fruits and seeds:* By far the most important factor that gave rise to the cultivation of plants for food is the food stored by plants in their fruits and seeds. Food may be stored as starch (e.g. cereals such as maize), or as protein (e.g. cowpeas and legumes generally), or in more complex forms such as spices (e.g. chilli, capsicum), or as sugars (e.g. mangoes, guavas, oranges) or as oils (e.g. coconut, oil palm, groundnuts).

*Respiration and Transpiration*

All living organisms, including plants, *respire* and give out wastes in the form of water vapour and carbon dioxide. Plants need oxygen for the breakdown of food (usually sugars or other carbohydrates) to release energy, and the carbon dioxide and water vapour produced during respiration are discharged through the stomata in the leaves. Respiration is the opposite process to photosynthesis (see Figure 3.2).

*Transpiration* is the process of losing water vapour mainly through the stomata in the leaves. In order to keep a plant turgid (i.e. with enough water in all its tissues), transpiration losses have to be constantly replenished through the roots. Therefore soil water must be sufficient for plant needs; otherwise the plant wilts and dies.

*Farm Cultural Activities*

Several cultural activities (pertaining to the raising of crops) are carried out largely to benefit the crop plants and ensure good plant growth. Some of these activities (e.g. weeding, erosion, transplanting) are very important indeed and are discussed to a limited extent in subsequent sections of this chapter.

## 3.2 PROPAGATION AND MANAGEMENT OF CROPS IN WEST AFRICA

The yield of crop plants is affected in many ways by four major factors: (a) the soil; (b) the climate; (c) the potential of the seed or planting material used; and (d) the farmer, especially in the way and manner in which he propagates and manages the farm and, in particular, the crop plants.

The first part of this section covers the kind and stages of propagation, while the second part is about cultural practices.

### 3.21 KINDS OF PROPAGATION

Plants can be *propagated*, or reproduced several times over, by using the *sexual technique*, in which the male gametes fertilise the female gametes and the resultant product (seed or fruit) is used for propagation (or planting). In other cases *asexual techniques* are used, in which a section of the living plant is induced to form separate roots and subsequently to develop into a separate plant.

Some of the important or commonly used asexual methods of propagation are the following:

(1) *Budding:* The *bud* or *scion* of a plant is successfully implanted on the stem of another plant of the same genus (the *stock*). For example, to overcome root diseases in citrus it is often the practice to use a resistant variety as the stock and to combine this with a high-yielding but poorly resistant variety on top for a high fruit yield.
(2) *Layering:* A part of the stem of a growing plant is anchored to the earth with a view to establishing a new root system at the point of contact between the bent stem and the earth. As soon as the new

plant is established, the connection with the old part is severed and the new plant is on its own.

(3) *Grafting:* This method of propagation though similar to budding is different from it. A twig cut with top buds and foliage is transferred from one plant to replace the top of another that has been cut off. This can be done with citrus, mangoes and most other fruit trees.

(4) *Cuttings:* The mature stem of the plant to be propagated is cut into smaller pieces, usually 15–25 cm (6–10 in) long, which show a number of buds that will subsequently give out branches and leaves of a new plant. The cut part is buried in the earth, at an angle away from the sun and with at least half the buds above ground level. This is the only effective way of propagating yams and tubers such as cassava.

(5) *Leaves:* These are sometimes used (e.g. cocoa). However, to make the method effective the leaves are previously treated with growth hormones to stimulate root formation as well as quick development of the bud cut with the leaves.

(6) *Rhizomes:* These are underground stems in which the plant can store some food. When cut into smaller lengths of 10–15 cm (4–6 in), they can be buried with scale leaves etc. and grow out as separate plants.

(7) *Runners:* These are the stems of crops such as grasses and sweet potatoes, which lie on the ground horizontally. These also can be cut into smaller pieces and partially buried to start new plants later.

(8) *Bulbs:* These are underground stems whose leaves are used as storage organs. The stem part is usually very short (or *telescoped*), and the fleshy leaves protect the terminal bud which will eventually grow into a new plant given the appropriate environment.

Sexual reproduction or propagation, discussed in section 3.14, involves the fusion, or coming together, of two gametes (male and female) to form a seed or fruit which is later planted and from which new plants emerge, given a suitable environment.

## 3.22 CULTURAL PRACTICES

What are cultural practices? The production of crops involves a number of activities and operations. Some of these operations are specifically carried out on crops at different stages of their growth and development. These activities (e.g. planting, weeding, watering, transplanting) are involved in the good growth and establishment of crops and are known as *cultural practices*. They are carried out with the main purpose of improving the growth and yield of the field crops. Cultural practices include pre-germination activities such as seed preparation, soil preparation and planting, as well as transplanting and post-planting activities.

*Seed Preparation*

Seed preparation involves activities like getting the right type of seed, cleaning them free of weed seeds and other impurities, as well as making sure that the seeds are viable. Viability is found out by carrying out a *germination trial.* To do this, the farmer has to plant 100 seeds in a nursery box under ideal conditions for germination; he then counts the total number of sprouting (germinating) seeds after the expected number of days needed for germination.

Carrying out a germination trial helps the farmer to know if his seeds are good enough for planting. It also helps him to know what his *seed rate* will be when he plants in the field. Seed rate (*SR*) is the amount of seed required to plant 1 acre or 1 ha (hectare). It is the product of the number of seeds to be planted per hole (*SH*) multiplied by the number of holes (or *stands*) in the acre or hectare (i.e. the *plant population, PP*). The plant population is calculated by dividing 1 acre or 1 ha by the *feeding area* per plant (*FA*). Thus, if *Sp*=spacing of holes and *Rg*=spacing of ridges or rows,

$$FA = Sp \times Rg$$

$$PP = \frac{1\ ha}{FA} \text{ or } \frac{1\ acre}{FA}$$

$$SR = PP \times SH$$

For instance, if maize is planted in holes 46 cm (18 in) apart on ridges spaced 91 cm (36 in) apart, with 2 seeds per hole,

$$\begin{aligned} FA &= 46 \times 91 \\ &= 4{,}186 \text{ cm}^2 \\ PP &= \frac{1 \text{ ha}}{4{,}186 \text{ cm}^2} \\ &= \frac{1 \times 10^8}{4{,}186} \\ &= 23{,}889 \text{ plants per hectare} \\ SR &= 23{,}889 \times 2 \\ &= 47{,}778 \text{ seeds per hectare} \end{aligned}$$

or

$$\begin{aligned} FA &= 18 \times 36 \\ &= 648 \text{ in}^2 \\ PP &= \frac{1 \text{ acre}}{648 \text{ in}^2} \\ &= \frac{43{,}560 \times 144}{648} \\ &= 9{,}680 \text{ plants per acre} \\ SR &= 9{,}680 \times 2 \\ &= 19{,}360 \text{ seeds per acre} \end{aligned}$$

However, seed rates are usually expressed in weights (pounds/kilograms/tons etc.) depending on the weight of each seed. A knowledge of the seed rate is essential so that the farmer can plan adequately for future planting and especially so that when he needs to buy the seeds or planting materials he can make adequate preparations.

In some cases, special treatments are necessary before the seeds can germinate. For one reason or another – because special efforts are needed to break the seed open, or the seeds are too fragile, or the young seedlings need shade or mulch, or the seeds are too expensive to risk being destroyed in the field – the pre-planting situation becomes inevitable. For instance, the oil palm kernel is very hard and tough, so the development of the embryo requires special treatment. Such cases require a nursery where special care and attention may be given to the seeds and seedlings before transplanting is carried out.

### *Soil Preparation*

Soil is a composite of heterogenous materials. Soil contains soil particles of different sizes, mineral particles, mineral salts, water, air, humus, and some living organisms (*soil microbes*) in varying sizes, amounts and proportions and in different combinations from time to time. For best yields of crops, it is essential for the soil to be prepared so as to give the plant the most ideal conditions for growth and development. To achieve this the land has to be cleared and the soil tilled and ploughed before the sowing of seeds or seedlings. Drainage of wet soil or irrigation of dry soil may also be necessary.

*Land clearing* is achieved in the following manner. First, using axe, machete and hoe, the undergrowth is cleared and small trees are cut, preferably in the dry season. Sometimes burning is done to reduce labour, even though burning encourages the destruction of soil microbes and topsoil minerals, especially if heavy rain follows the burning. However, in one form or another clearing is essential, and the usual practice by peasant farmers in most parts of Africa is to leave the heavier, bigger or more economically useful trees (e.g. palms, fruit trees, exportable timber). Stumping is also avoided because of the labour cost. But if cattle or tractor ploughing as opposed to hilling (making hills) is to be done subsequently, stumping *must* be thoroughly carried out to avoid damaging the plough or the animal or machine pulling the plough. This is very important.

*Tilling* helps to turn the soil over, thereby improving soil structure. It also helps to destroy the larvae and pupae of pests by exposing them to the sun. If manure or fertiliser is to be used, this can be buried in the soil in the same operation. Tilling also cleans the soil by exposing roots to heat, drying and rot and buries the leaves on top.

It is important for the tillage depth to be shallow so as not to bring the lower subsoil up to the surface. A depth of 10–15 cm (4–6 in) is considered shallow, while a 15–20 cm (6–8 in) depth is considered normal. The appropriate depth to be used will also depend on the nature of the soil and the crop to be planted. Where the topsoil is thin it is advisable to do only shallow tilling.

*Ploughing* may follow after tilling. If manure is buried it should be allowed to settle before ploughing. In this case a period of 2–3 weeks after tilling is considered adequate, but the longer the period of rest, the better.

Ploughing requires extra care and special equipment, such as a plough or ridger with a man (farmer), a horse, cattle or a tractor drawing the plough. If the soil is very light, simple tines mounted on beams may be sufficient and ploughing may not be necessary. If a sloping surface is to be ploughed, it may be advisable and sometimes necessary (depending on the degree of slope) to arrange the ridges across the slope and use *contour banks*. This is called *contour ridging*. It is used to check erosion and prevent the excessive loss of rainwater. After ploughing, the field is ready for sowing or transplanting as the case may be. It is important to wait for the rains before sowing or transplanting.

Where the soil has too much water (is *waterlogged*) or where the flow of subterranean (underground) water is poor, the need is obvious for carrying out *drainage* before any successful planting can be achieved. The aim of drainage is to remove excess water from the soil and allow in more air for a balanced growth of plant roots and the crop plants as well.

Two important and common methods of drainage are: (a) the use of subsurface drains or pipe drains; and (b) open-surface drains or gutters. Open surface drains are used more commonly since they are cheaper, as pipe drains break too easily, particularly in the tropics. Where open drains or ditches are used, care must be taken to ensure that the slope is sufficient to prevent their rapid blocking due to the deposition of silt (*silting*). At the same time it is necessary to avoid too great a slope on which rapid water flow may give rise to erosion. To check erosion, *bunds* (small banks) are built across drains, and the bunds collect silt and conserve water.

*Irrigation* can be defined as the controlled application of water over a land area. This may be carried out when there is need to supplement insufficient rainfall in the area or at the time of planting if water or rain in sufficient amount is not available.

Several irrigation methods are available, depending on (a) available funds; (b) land area to be irrigated; (c) the water supply for the irrigation; (d) soil type; and (e) the crop to be irrigated (seasonal water requirement and market value of the crop).

The most common methods used in Africa are:

(1) *Flooding method:* A dam is erected across an all-season river or

interconnecting ditches from a big river are used. This is only possible where there is an abundant water supply nearby.

(2) *Furrow method:* In this method, applicable in the case of, say, onions, potatoes, sugar cane etc. which are planted in furrows, the furrows serve as ditches for letting in water from a reservoir, pipe or well.

(3) *Sprinkling method:* The water comes through a pump which forces it out of the pipe through a narrow nozzle or outlet. The nozzle creates a spray or sprinkling effect, which throws the water to quite a distance round the pipe outlet. Sprinkling is very effective with lawns and broadcasted crops (e.g. flower beds), particularly where the soil has a high infiltration rate. It is also useful where there is a great risk of erosion if other methods are used.

*Planting and Transplanting*

Seeds of crop plants are planted in fields (e.g. maize, cowpeas) or sometimes in *seed beds* preparatory to *transplanting* (e.g. tomatoes, vegetables, tobacco, rice). Where the seeds have to be planted in a special seed bed, usually referred to as a *nursery*, care is taken to ensure good germination.

The planting operation involves having viable seeds which are buried under the ground, the depth varying according to the size of the seeds. Water must be supplied. Sometimes chemicals to destroy pests, as well as fertiliser to accelerate growth, are mixed with the seeds at planting.

The other method of planting, called *broadcasting*, involves spreading or throwing in arc waves handfuls of the seeds over the area in a steady sequence. This is usually done with grass and rice seeds. After the seeds are broadcasted, watering takes place immediately, to give a partial earth cover for the seeds and especially to prevent the seeds from being destroyed or eaten by insects, birds and other pests.

Transplanting may be necessary for various reasons:

(a) to increase root growth and tillage by the emergence of offshoots (e.g. rice); (b) to improve yield (e.g. tobacco); (c) to raise a stock for subsequent vegetative propagation (e.g. for budding); and (d) to supply the plant with the most suitable soil conditions (e.g. fruit trees, permanent crops).

Where transplanting has to be done, the young *seedlings* raised previously under ideal conditions in the nursery have to be moved carefully so as not to disturb the root system more than necessary. Field planting of seedlings requires specific spacings, pit dimensions and other cultural operations aimed at guaranteeing the good growth and development of the new plants. It must be ensured that the new environment is ideal, with good topsoil, fertiliser where necessary and enough moisture.

*Post-planting Practices*

There are three sets of post-planting cultural practices: (a) watering, mulching, trimming and general maintenance practices to keep the plant in good condition during growth; (b) weeding, to reduce competition from extraneous plants; and (c) harvesting operations, to ensure that all that is produced is protected, gathered and well stored for later sale or consumption.

*Watering* is done: (a) to provide adequate moisture for the plant; (b) to control the soil temperature; (c) to supply adequate nutrition to the plant; and (d) to increase plant turgidity, thus preventing *lodging* and wilting.

*Mulching* is also done for a number of reasons: (a) to control soil temperature; (b) to conserve soil moisture; (c) to protect the roots of this young seedling; (d) when the mulch rots, to produce some humus and improve soil structure; (e) to protect the planted sets or seeds from direct heat, thus checking drying out (e.g. yam tubers); and (f) to reduce the action of insect pests that may attack the growing plant or seedling.

In mulching it is important to use the *clean*, *dry*, *dead*, *vegetative* parts of plants. Seeds of weeds can be easily transplanted if the mulch used is not clean. And if it is not completely dry, mulch may start to rot and create heavy bacterial action, which may heat up the soil and do great damage to the roots and other parts of the young growing plants.

*Weeding* is the practice of removing weeds from the farm. Weeds are growing plants that were not planted and are not wanted where they are growing. Hence they have to be removed, for a number of reasons: (a) to allow enough food and sunlight for the crop plants that are wanted; (b) to avoid impurities in the seeds that are to be harvested; (c) to prevent known secondary hosts of some pests and other disease-causing organisms (e.g. bacteria, fungi, viruses); (d) to avoid poisoning, resulting from animals grazing on poisonous weeds in a pasture; (e) to avoid damaging fieldworkers; and (f) to beautify the surrounding, particularly in a flower bed, lawn or fruit orchard.

Weed control is the cultural practice by which weeds are removed. The basic policy in any cultural practice is to remove the weeds before they drop seeds. This is done by one or a combination of the following methods:

(1) *Mechanical control:* A machete, a hoe or machines are used either to cut off the vegetative part (e.g. hedges) or for the complete

uprooting and drying (or *rotting*) of the weeds. These methods cost labour and/or fuel and can be expensive, depending on the frequency of weeding, extent of weed, type of weed and soil type.

(2) *Biological control:* Cover crops, legumes and grasses (lawns) have a versatile (rapid) growth rate, and because of their prostrate or twining stems they grow horizontally to the ground and are in a position to cover the soil and prevent weeds by choking or smothering them out. In some cases, browsing animals (e.g. goats, rabbits) feed on such weed-bound areas. In some orchards and plantations, grass lawns are kept low and weed free through intergrazing with livestock (e.g. cattle, horses, sheep, goats).

(3) *Cultural control:* Intercropping, rotational planting, rotational grazing of pasture and adequate tillage are all cultural practices that are cheap and effective in the control of weeds if properly followed.

(4) *Chemical control:* Chemical preparations (e.g. herbicides, weedkillers) can be applied or sprayed on weeds. Selective weedkillers have been developed and are marketed; these can destroy special types of plants. Sometimes the excessive or indiscriminate use of fertilisers, and their application at the wrong time, encourage weeds. Care should be taken in the use of herbicides, as some of them are poisonous and can prove harmful to other plants, grazing animals or the humans applying the herbicides.

The last field operation with crops is *harvesting*, while the major activity at the start of the post-harvesting operation is usually *storage*. Harvesting is the culmination of the farming operations and should be given as much care as possible to avoid losses in the field. The important steps to watch in harvesting and storage are:

(1) Harvesting should be timely – not too soon and not too late, with a mature and relatively dry crop and with suitable weather conditions.
(2) Harvesting should be complete and clean – the farmer should remove as much as he possibly can of the produce that is ready.
(3) The harvest should be kept in good condition by good storage in adequate insect-proof dry containers.
(4) Where the seeds or fruits are not dry, a drying process to reduce the water content and minimise mould and fungal attack is useful and must be adequately carried out.
(5) Storage should be effective against rain, excessive direct heat, thieves, insects and other pests (e.g. birds, rats); other areas of losses should be equally minimised.

## 3.3 CROP IMPROVEMENT AND PROTECTION

### 3.31 CROP IMPROVEMENT

*Aims of Crop Improvement*

The aims of crop improvements are: (a) to increase the yield of the crop; (b) to improve the disease resistance of a particular strain; (c) to adapt the crop to some ecological area; (d) to increase the nutritional value of a particular crop; (e) to introduce some new desirable and quality improvement character(s) into an existing crop (e.g. uniform ripening in pineapples, fruits or even tomatoes); (f) to improve the uniformity of planting material, thus maintaining a strain that can later be propagated extensively; and (g) to increase scientific interest and knowledge.

The cost of crop improvement is high. It involves specially trained staff working with expensive equipment for a number of years. Hence, because of costs, personnel and necessary regulations, the responsibility for improving and for the control and distribution of improved genetic material (e.g. seeds) is solely the government's.

All crops that are cultivated are eventually for human and animal consumption. The aim of any crop improvement from the national point of view is ultimately to increase and improve the capacity of the crop to satisfy this demand fully, with some surplus wherever feasible.

*Methods of Crop Improvement*

The three most common methods for achieving the above objectives are by the processes of introduction, selection and cross-breeding.

The *introduction of new crops, new varieties or new types* either by accident or deliberately is one of the oldest and probably the cheapest method of improving the crops of a country. Over the years a number of important crops have been introduced into West Africa (e.g. coffee, tea, cocoa, tobacco, cassava, maize). As recently as a decade ago a new type of fibre crop, kenaf, was introduced into Nigeria, and some exotic temperate vegetables (e.g. lettuce, cauliflower, cabbage) have also been successfully introduced into West Africa.

Introduction of a new crop involves bringing the reproductive material (e.g. bulbs, seeds, cuttings, seedlings) into the country, of course with the approval of the government. Before any new plant is introduced the material is first quarantined, i.e. kept away from contact with other plants or animals for detailed observation and examination.

Quarantine is necessary for several reasons, such as: (a) to prevent the introduction of poisonous or dangerous materials; (b) to protect existing varieties or crop types from being genetically polluted; (c) to prevent the importation of foreign diseases and infections; (d) for record purposes and for full information on the material being introduced; and (e) to preserve the new material from being attacked or destroyed through local infection.

New crops or varieties are introduced depending on their potentials and characteristics. The introduced crops are subsequently established as a *pure strain*, multiplied or increased in number, and may be distributed for general planting. Sometimes they are crossed with local breeds or types to produce *hybrids* and new strains. The scientists who carry out this type of work are known as *plant breeders*.

Crops can also be improved from local breeds by picking out only the plants that indicate certain desirable characteristics. Those identified are then developed and multiplied for general distribution. The scientific process of identifying, picking out and establishing such types or varieties with the desirable characteristics from existing local plants is known as *selection*.

The scientific principle in selection is the hereditary potential of certain characters passed on through *genes* during the cell division processes of *mitosis* and *meiosis*. The desirable characters are first observed, their genetical elements are identified, and when the selected plants are reproduced they transmit these characters to their offspring.

It is important to separate two types of selection. One is *mass selection*, when several plants showing certain desirable characteristics are selected and their seeds or cuttings are collected and subsequently planted. There is also *individual plant selection*, by which only one plant is ultimately picked (selected) for future propagation, multiplication and mass distribution. These two methods complement one another.

Selection is usually a time-consuming operation. Depending on the vegetative life of the plant and the nature of the characters being observed, the selection process may last from five to fifteen years. A lot of money is also required in the process for the laboratory and fieldwork required. Of course, any type of crop improvement is expensive, but it is an expense that is always rewarding when the results prove successful.

The third method of crop improvement is *cross-breeding*. When two pure but different breeds (called *parents*) are cross-fertilised, the product is known as a hybrid. The process of maintaining the hybrid with the desirable characters from the two parents is known as cross-breeding.

Cross-breeding entails a lot of work, time and expense, since the two parents have to be maintained in their pure strains to avoid contamination. Two advantages of cross-breeding are: (a) it develops hybrid vigour, i.e. strength in plant and productivity; and (b) it allows for the simultaneous combination of a number of desirable characters (e.g. increased yield, improved quality of produce, uniform ripening time, palatability) in two species or varieties, an advantage which often is not possible with other methods of crop improvement.

*Improving Crop Yields*

Crop improvement is easily confused with the improvement of crop *yield* or output. Even though crops can be potentially improved for certain desirable characters through methods of crop improvement, the cultural practices have to be improved before these improved crops produce their best or maximum yields. By improving cultural practices the full potential of all characters of all improved crops are realised. Plants will grow to the best of their ability, produce their best fruits or seeds and have the best overall performance.

The important cultural practices that effectively bring out the full potential of improved crop plants are the following:

(1) *Planting:* Planting must be done correctly (depth), be timely and have enough moisture and warmth. Pre-planting cultivation has to be well done.
(2) *Plant population:* This must be just enough, not too thin or too thick, with spacing as recommended for the improved variety.
(3) *Organic manure and fertilisers:* These must be the correct ones for the crop and soil type and must be correctly applied, i.e. at the right time and in the right quantity.
(4) *Weeding and other cultivation:* This must be done properly and at the right time as far as weather permits.
(5) *Pests and diseases:* If not properly looked after and controlled, pests and diseases can lead to a total loss of crops.
(6) *Farm organisation and planning:* These indirectly affect total farm operations. Care should be taken to have expert advice when necessary, as this can mean a difference between failure and success in farming.
(7) *Crop rotation:* The use of crop rotation and improved husbandry generally improve yield while keeping the farm and soil at peak production.

### 3.32 PESTS, DISEASES AND CROP PROTECTION

A *diseased condition* can be defined as an abnormality or malfunctioning which is unwholesome or unnatural in any part or the whole of a plant. This situation usually results from an infection or attack by other bodies or from some internal dislocation or malfunctioning within the plant itself. A diseased condition therefore is different from a *pest attack* in which the plant or crop product indicates visible signs of damage resulting from the attack. It is also important to remember that in some cases a pest attack may be also followed by an infection, usually referred to as *secondary infection.* For example, the cotton bollworm is a beetle pest which feeds on the juice of the cotton fruit and subsequently gives room for fungi attack which may destroy the whole boll.

Similarly, an infection by a disease causal organism may lead to subsequent pest attack. For example, wilting or nutritional disease resulting from loss of moisture may lead to lodging of the plant, and subsequently the plant buds and leaves are eaten by caterpillars or rodents.

Pests attack crop plants at two main stages: in the field and in the store. Some pests, particularly birds, attack in the field right from the planting of seeds, through seedling stage up to the fruit formation. There are a great many of them, and the important ones are mentioned briefly for each crop in Chapter 4.

### *Major Crop Pests*

The major crop pests are classified for convenience into the following groups:

(1) *Rodents:* These feed on all parts of plants but mostly attack cereals, tubers and seedlings. Monkeys especially feed on fruits and bananas. Other animals (e.g. goats) feed on the herbage of several crop plants.

(2) *Birds:* These feed mostly on grains, berries, and soft-coated fruits.

(3) *Insects:* As larvae they feed on leaves and flower shoots; as adults, on grains both in the field and in the store. Most insects, particularly beetles, suck on the growing parts of the plants as well as the flowering shoots, causing physical damage. In addition, sometimes their attack opens the way for secondary infection from fungi, bacteria, viruses and other disease-causing organisms.

(4) *Mites* (arachnids): These are very small animals with mouth parts that are adapted for piercing (boring a hole) and sucking plant juice. They feed on virtually all vegetative parts of crop plants.

(5) *Nematodes* (worms): This group (e.g. eelworm, roundworm) can exist both as free-living forms and as parasites in the soil. Some of them attack the roots of crop plants such as legumes, tomatoes and other garden crops, causing great damage and heavy losses to farmers.

Some of the more common insect pests of crop plants in West Africa include crickets, locusts, grasshoppers, termites, thrips, aphids, butterflies, moths, flies and beetles. The plants they attack and the nature of their damage is given in Table 3.2.

Table 3.2 *Crop Pests: Their Damage and Control.*

| *Crop Pest* | *Crops Attacked* | *Nature of Damage* | *Control* |
|---|---|---|---|
| Locust | Cereal crops, legumes, pasture. | At migratory phase they swarm over large areas, stripping crops bare of any foliage (or vegetation). | Destroy *all* locust-breeding centres at solitary phase. |
| Grasshopper (*Zonocerus variegatus* (L.)) | Cassava, cereals, legumes, citrus, coffee. | Leaf-eating pest which devastates foliage in dry season. | Hand pick; spray with insecticide; practise clean husbandry. |
| Cricket | Various crop seedlings. | Eat up seedlings. | Seal up cricket holes; protect important seedlings with collars. |
| Aphids | Citrus, groundnut, cotton, beans. | (a) Suck plant sap from buds and leaves.<br>(b) Vector carriers of viruses, e.g. rosette on groundnuts.<br>(c) Retard growth of young plants. | Spray with insecticide. |
| Thrips | Coffee, banana, tea, sometimes the flowers of several shrubs. | Feed on sap after piercing through like aphids. | Spray regularly. |
| Butterflies and moths (Lepidoptera) | Caterpillars eat virtually any crop leaves: e.g. army worm on maize; bollworms on cotton; larvae and pupae of moth on cocoa in store; cocoa trunk-borer moth; yam leaves. | (a) Adults feed on nectar and help pollination, but larvae (caterpillars) defoliate plants.<br>(b) Damage (taint) cotton lint.<br>(c) Carry fungal and virus diseases. | Hand pick; spray with insecticide; use crop rotation and cultivation; destroy pupae in the soil. Fumigate stored cocoa; inspect regularly and store adequately. |

Table 3.2—*continued.*

| *Crop Pest* | *Crops Attacked* | *Nature of Damage* | *Control* |
|---|---|---|---|
| Butterflies and moths (continued) | | (d) *Buseola fuss* (stem-borer) on all cereals. (e) Moths feed on cocoa trunks and pod husks. | |
| Weevils (Coleoptera) | Grains of all types, groundnut, cashew, banana, kola, coffee, sweet potato, beans/legumes (bruchids). | Eat grains, in store mostly; larvae burrow into plant stems, leaves, pods and fruits; roots of sweet potato and banana eaten and destroyed. | Hand pick; store well; spray or dust with insecticide; sterilise soil. |
| Beetles (Coleoptera) (e.g. rhinoceros beetle) | Yam, beans, cucumber, melon, vegetable marrow, okra, coconut. | Larvae and adults feed on tubers; adults feed on many crop leaves; leaves perforated by adults while feeding; terminal buds and stem (e.g. coconut) eaten. | Hand pick; sterilise soil; use crop rotation; practise clean husbandry; spray or dust with insecticide. |
| Nematodes (e.g. eelworms, roundworms) | Tomato, legumes, potato, carrot etc., coffee. | All parts of plant attacked and eaten; burrow in stems; roots destroyed; e.g. root-knot nematodes causing wilting, retarded growth and low yields. | Sterilise soil; practise deep cultivation, crop rotation and clean husbandry. |
| Birds (e.g. *Quelaquela aethiopica*) | Sorghums, maize, rice, cereals. | Eat grains; suck milk of fruits; nuisance effect. | Destroy bird colonies; use scarecrow; do timely harvest; shoot. |

disease in cocoa, root-knot in pawpaw and fruit rot in citrus. Damping-off disease in plant seedlings is caused by another member called *Pythium* sp.

Control is by using resistant varieties, and spraying with copper mixtures helps. Avoid excess moisture. With cocoa, avoid too thick undergrowth. Clean husbandry and regular routine spraying reduce infection.

Popularly known as sac fungi, *ascomycetes fungi* include the useful penicillium, but a good number of them cause devastating plant diseases. *Erysiphe* spp. cause powdery mildews. They feed on their hosts through their absorbing organs, called *hausteria. Claviceps* sp., another member, is a parasitic fungi on cereals and grasses, infecting their young flowers right through to the ovaries until finally a (*sclevotium*) poisonous, dark-coloured, dormant mycelium called *ergot* replaces the grain.

In West Africa the two most important and dangerous sorts of *basidiomycetes fungi* are smuts and rusts.

(1) *Smuts:* These fungi (e.g. *Ustilagihales* spp.) affect all cereal crops all over the world. On sorghum, the heads affected are light with the grains replaced by greyish cylindrical spore sacs filled with black spores. Other clean seeds are contaminated easily, and if these are planted the disease spreads. Several smuts are known – grain smuts, open smut, head smut and long smut, caused by different *Sphaceolothera* spp. Smut on maize is caused by *Ustilagomaydis* spp., which affect the grains, leaves, stems and flower, producing black 'boils' of black spores. On rice, leaf smut is caused by *Entyloma oryzae*, which produces numerous small black spots on the leaves. Control is by roguing: uproot infected plants and bury, or burn and bury, before the spores burst. Strict crop rotation is necessary, and the practice of clean husbandry helps considerably.

(2) *Rusts:* These are caused by the obligate parasitic fungi belonging to genus *Puccinia*, have worldwide distribution and attack a wide variety of cereals (e.g. maize, rice, wheat) and grasses all over Africa. They give rise to loss of chlorophyll and poor, if any, grain harvest. Control is by spraying with copper fungicides (sometimes ineffective), planting resistant varieties and practising crop rotation.

*Fungi imperfecti* are regarded as imperfect because they reproduce only asexually. *Penicillia* sp. is a useful member, but a number of the others cause a lot of diseases. The more important ones in West Africa are *Fusarium* sp., causing Panama disease on banana in the Cameroons and Nigeria, and *Alternaria* sp., causing coffee berry disease in Nigeria, Ivory Coast etc.

*Virus Diseases:* The two main categories (bacterial and fungal) of disease-causing organisms, discussed above, are easily identified either by the naked eye or through a microscope. Viruses, on the other hand, are too small even for ultra-microscopes and can only be established by their biological activity. Apart from being very small, viruses can survive only in other living cells and cause severe pathological conditions in their hosts.

Viruses are easily transmitted through the soil, by insect vectors such as aphids and ants, by contact and by seeds.

The more common virus infections in West Africa include the following:

(1) *Cassava mosaic:* Leaves are distorted, with pale-green streaky pattern. Plant growth is stunted. The white fly (*Bemisia* sp.) is a vector of this virus. Control is by avoiding the use of infected cuttings, by using resistant varieties and by the control of white fly breeding. Uprooting infected plants reduces spread.

(2) *Groundnut rosette:* The virus is carried by aphids. Attack can be very serious, leading to a complete failure and total loss of the crop. Symptoms on the plant include stunted growth, with marked mosaic on leaves; sometimes some chlorotic marks in leaf veins precede heavier attack. Control of aphids by picking and spraying and raising healthy plants minimise aphid attack and indirectly reduce virus infection. Use of resistant varieties and early planting are highly recommended preventive measures.

(3) *Tobacco leaf curl:* This is caused by another group of viruses again usually transmitted by the white fly. Leaves of tobacco become distorted and curl up, veins become highly thickened and leaf outgrowths may be formed. Crop loss is heavy as damaged leaves are normally not harvested but uprooted with the plant and burnt. Control is by using resistant varieties, roguing (uprooting) tobacco plants at the end of the cropping season, avoiding the use of contaminated seeds and spraying against white fly. Crop rotation should be practised.

(4) *Tobacco mosaic:* This virus disease, easily transmitted by contact or touch, can assume epidemic dimensions if not properly controlled. Mottling of leaves is very common, and in young leaves the veins look transparent. Planting resistant varieties helps to reduce losses from infection.

There are a number of other virus infections (e.g. maize streak, citrus tristeza, potato mosaic) which are important but not very common in West Africa. A number of viruses also attack tomato crops. In all cases, virus diseases are characterised by mottled and distorted leaves with stunted growth and lesions of different colours. It is also important to bear in mind that virus infection often makes the plant susceptible to (other) secondary infections which will also appear on the crop plant.

*Disease Prevention and Control*

The prevention of plant diseases is much easier than their control. Once the crop plants are attacked, control is always difficult, expensive and often ineffective, especially if delayed. Disease-causing organisms are normally present at all times everywhere, but when unchecked their population increases beyond the level of ecological balance, thus causing great losses to farm crops. The aim of all preventive measures is therefore to keep the population of disease-causing organisms to a very low level.

*Prevention* can be achieved by: (a) practising clean husbandry, i.e. keeping plants healthy and farms clean; (b) regular weeding and the clearing of undergrowth; (c) the use of resistant varieties; and (d) the rotation of crops and rotational grazing, which keep the level of infectious spores and organisms relatively low.

*Control* methods include: (a) the immediate application of chemicals by dusting or spraying, which if properly carried out will at least minimise the spread of disease; (b) avoiding contact with healthy plants after touching infected plants, as virus infection can easily be spread by contact; (c) roguing, uprooting and burning infected plants; and (d) reporting infection immediately to an agricultural officer for immediate control action.

*Chapter 4*

# Notes on Some Important Crops of West Africa

## 4.1 ANNUAL CROPS

Crops that grow, reproduce and die within one calendar year or one cropping season are known as *annual* crops. Those that survive for more than a calendar year or produce for two or three cropping seasons are known as *perennials*. Others (e.g. rubber, cocoa, oil palm, citrus) tend to last for over twenty to thirty years and are commonly known as *permanent* crops.

These three main seasonally classified farm crops require several agricultural operations, which are similar, but a few unique features have also to be known. Hence the discussions that follow are based on the cultural requirements of the different crops.

For convenience, the annual crops have been grouped into the following:

(1) *Tubers* (e.g. cassava (root tuber), yam (stem tuber)) produce heavy fibrous starchy stored food underground in their roots or stems respectively.
(2) *Cereals* (e.g. rice, maize, sorghum, wheat) are some of the grasses cultivated for their seeds which are consumed by man.
(3) *Legume crops* (e.g. beans, peas, groundnuts) have semi-erect or creeping stems and produce seeds in pods either above ground (e.g. beans, peas) or underground (e.g. groundnuts). Legume seeds are rich in protein and oil. Furthermore, their roots produce *root nodules* in which live nitrogen-fixing bacteria which indirectly improve the nitrogen content of the soil.
(4) *Industrial crops* (e.g. tobacco, chillies, cotton) are cultivated mainly for their industrial uses. Tobacco, used in the making of cigarettes, cigars, pipe tobacco and snuff, is of great commercial interest and brings in a substantial income to peasant farmers.
(5) *Vegetable crops* are grown mainly in vegetable gardens where they are given special care and attention, such as heavily manured beds, regular (daily) water and daily cultivation. Some are produced for their leaves (e.g. lettuce, amaranthus), others for their root storage organs (e.g. carrot, radish, ginger) and others for their fruits (e.g. tomatoes, egg plant, melon).

## 4.11 TUBERS

Yam (*Dioscorea* spp.)
Other names: Yoruba – *isu*; Igbo – *ji*; Fanti (Ghana) – *edwiw*; Sierra Leone – *toli.*

Yam is a stem tuber crop of the tropic and semitropical belt. There are well over ten different varieties known, cultivated and eaten in different forms (boiled, roasted, fried, pounded) all over West Africa.

It is a major staple of West Africa. The peel (skin) from the tuber is fed to livestock. The three most common varieties are the white yam (*D. rotundata*), yellow yam (*D. cayenensis*) and water yam (*D. alata*).

*Land Preparation and Planting:* Traditionally, the best soil after bush clearing is reserved for yam since it is a major staple and a heavy feeder. Heaps and mounds 0·9–1·2 m (3–4 ft) apart are prepared well in advance of the rain; 0·9–1·2 m (3–4 ft) ridges are also effective. Planting material is either yam seeds raised in a nursery the previous year, or yam sets cut out of a bigger tuber also of the previous harvest.

Time of planting varies depending on climate (the end of the rainy season is used for clearing the forest), variety, tradition and the rains. October to March the following year is usually the yam-planting period. Seed rate is roughly 2·6–5·1 tonnes/ha (1–2 tons/acre).

The tuber, or seed yam weighing 0·5–0·8 kg (1·1–1·8 lb), is buried in the heap/mound/ridge and sometimes covered with mulch, especially if planting is done in October–December. Seed yams or tubers being planted must show 'eyes' or buds. Immediately after germination, staking is done, and the young vines are trained to twine round the sticks. Staking is done carefully to prevent delayed growth, which may be caused by any damage to the growing tip of the plant. It is essential to be watchful to replace seeds or tubers that fail to germinate.

Apart from staking, the only other cultural operation during growth is weeding. In terms of labour and seeds, yam is a very expensive crop to produce. To make it worthwhile, therefore, weeding has to be done regularly to keep down weeds as well as to keep the heap, mound or ridge at a high level and covered with a lot of topsoil. This is done to ensure the highest possible yields, without which planting of yam will result in a net loss to the farmer and a very expensive crop on urban markets.

Fertiliser application, especially nitrogen, is recommended for increased yield. Some response to the low level application of potassium is also good. Fertilisers should never be applied before the rains, and even then they should be applied at recommended rates to prevent excessive rotting of the tubers after harvest.

*Harvesting and Storage:* Yams mature in 8–10 months after planting, and harvesting is indicated by the browning and drying of the vines. A very luxuriant vegetative growth of 8–10 months indicates a very good harvest.

Harvesting is by hand and hoe or machete, which is used in digging out the tuber very carefully to avoid bruises. Bruises open the skin to fungi infection and cause drying out, loss of quality and a lower market price. Harvesting stretches for 2–3 months between October and January depending on tradition, available storage space, the state of the vines and labour availability. Early harvesting in July–August is possible in very wet areas.

Harvesting is labour-intensive because of the care needed in harvesting and the weight of tuber to be carried from the farm to the storage barn. Yield is 10·2–15·3 tonnes/ha (4–6 tons/acre). Storage in the traditional way is done by tying tubers to upright sticks in the covered barn. Tubers are also stored in big holes dug out of the ground and covered with earth. This is only possible in areas with free drainage so that the pit is not waterlogged. Cold room storage is being tried in research stations but is still too expensive for general application.

Yam is sometimes better stored as either sun-dried yam or yam flour; both forms are easily reprocessed into food after storage, using boiling water.

*Diseases and Pests:* The most important diseases of yam in West Africa are yam mosaic and witch broom.

*Yam mosaic* is a virus disease giving rise to stunted leaf growth, leaf curl and distortion of veins showing chlorosis. It results in a very poor yield.

*Witch broom* is caused by the fungus *Phylleutypa dioscorea.* The white yam variety so affected has stunted growth with chlorotic leaves and poor branching. Long brown streaks of the fungus can sometimes be seen bursting through the stem cortex.

Control of the diseases generally are as discussed in section 3.32. Use resistant seeds and practise crop rotation. Generally, for virus attack, uproot and burn infected plants. For fungus attack, cut off and burn infected parts. Fungicide may be sprayed on infected stands. Clean husbandry is also helpful.

Several pests attack yam. The tuber is affected by the larvae and adult of the *yam tuber beetle* (*Heteroligus meles* (Billb)), which eats up most of the tuber. The larvae burrow into and voraciously eat away the tuber. Control is by spraying with Aldrin and practising crop rotation.

Yam shoot is affected by the insect *Ptyellus grossus*, which feeds on (sucks) young shoots causing yellowing of vines. Control by hand picking is sometimes effective.

Leaves are also food for the caterpillars and adults of *Cnoceris livida* (Family Lepidoptera) which occur in large numbers after the rains.

*Eelworm* attack yam in the soil and store, giving rise directly to yam rot and secondary infection by fungi as well as attack by other pests.

*Yam scale* is caused by *Aspidiclla hartii*, which feeds on yam mainly in the store and to a lesser extent in the field. Damage is reportedly very high in the dry season.

Control of pests is by spraying or dusting with insecticides and the practice of crop rotation. Hand picking is also helpful.

Cassava (*Manihot utilissima* Poli; *M. esculenta*)

Other names: Yoruba – *gbaguda*; Hausa – *rogo*; Liberia – *bai*.

Cassava, a native of Brazil (manioc), is a root tuber. The stored starch in the roots provides adequate calories in the food of a number of Africans in West Africa. The popularity of cassava on peasant farms depends on five important factors:

(1) Cassava, once established, survives adequately and is thus usually planted at the end of the traditional rotation just before the farmland is allowed to grow into the well-known secondary bush fallow.
(2) Once the cassava crop is mature, usually after 9 months, harvesting can be easily spaced out at any time during the year. Harvesting operations are therefore carried out at convenience when the cassava is required or when there is available labour for harvesting and processing. The cassava roots build up their prussic acid (HCN) level if stored after harvesting for more than 24 hours.
(3) Cassava is reputed to be capable of successful growth under a wide range of soil conditions and even on soils with relatively low levels of soil nutrients. This international adaptability explains the phenomenal widespread cultivation and use of the crop through the tropical rain belt of West Africa.
(4) Cuttings are a very cheap way of propagation. It is also very easy to introduce and multiply the good strains of the crop. The cost of production is on the whole very small.
(5) The use of cassava is very extensive, and nearly all parts of the plant are useful. The root when harvested is grated, fermented and fried into granules, or ground into flour which is reconstituted with boiling water into food. The starch is extracted and used in the cloth dye industry or made into foods such as tapioca and others. The young leaves are useful for vegetable stew or soup, while the mature leaves as well as root peels are used for livestock feed. Fresh cuttings from the stem are used for replanting, while the dried (or dead) stem is a cheap source of low heat firewood.

Two main groups, the *sweet* and the *bitter* cassava, have a wide area of cultivation in Africa. The best soil is a rich, freely well-drained, sandy loam. In too rich soils, excessive vegetative growth may reduce root storage. Cassava can survive prolonged drought but at high temperatures.

Cassava as food has a trace (less than 1 per cent) of protein, and therefore as a food it needs to be eaten with other high protein food substances. The presence of cyanogenic glucosides is an important deterrent to its use when raw, as these glucosides break down to form a poisonous product – prussic acid. The bitter variety of cassava contains more of this acid than the sweet one.

*Land Preparation and Planting:* Normally, cassava is a second or interplanted crop in which case the preparation for the first crop will usually be enough. It is also possible to grow cassava on the flat with only a little cultivation. Planting is best done at the beginning of the rains, when cuttings about 24–30 cm (9·5–11·5 in) in length, depending on the variety, are buried for half to three-quarters of their length. The best cuttings to plant are those from the middle portion of the plant. They should be planted at an angle of 45° to the ground, facing away from the sun.

Two important cultural operations essential to the good growth of cassava are weeding and crop protection. Two weedings at the early growth stage will be enough. Crop protection against pests (especially grasshoppers) and caterpillars is vital. Fertiliser application is known to increase yield on very poor soil, but this should only be applied after the crop is fully established.

*Harvesting and Processing:* Once cassava is matured, usually after 9 months, it can be harvested at any time as required. It is a productive crop producing yields of 10–20 tonnes/ha depending on variety, soil condition and age at harvest. Yields of 25 tonnes/ha have been recorded on experimental stations.

Processing has to follow harvesting since cassava, as explained previously, cannot be stored. Processing is done traditionally. It involves peeling the root, then grating and fermenting before frying to make cassava granules called *gari* (Yoruba). The root after drying can also be pounded into cassava flour. Tapioca and cassava starch, are also useful traditional preparations from the cassava root.

Cassava peels and cassava factory wastes are useful livestock feeds. Cassava leaves, which are rich in protein, oil, minerals and vitamin C, are harvested from time to time and cooked into stew for food.

*Diseases and Pests: Cassava mosaic* (virus) is a common disease of cassava. The symptoms include distorted leaves with a marked pale green pattern, stunted plant growth and poor tuber yield. The virus is widespread, carried by the white fly. Control is by the use of resistant varieties and by uprooting infected stands.

*Brown spot* is caused by the fungus *Cercospora lenningsii* Allesch. Spots appear as irregular brown marks on the leaves. White spots can also be seen, but these are symptoms due to the attack of another *Cercospora* fungus.

*White thread* (*Fomes lignosus*) is another widespread disease of cassava. Losses are high, sometimes as high as 25 per cent of expected yield. The roots (tubers) are covered by threads of this soil fungus, and rotting of tubers sets in. Control is by crop rotation, using resistant varieties and harvesting the roots early.

The major pests include *grasshoppers* (*Zonocerus variegatus*), which feed on the leaves and can only be controlled by hand picking and spraying with chemicals.

Several species of leaf-eating *beetles, bugs* and *goats* also do great damage to cassava leaves and young stems. The *cave rat* also eats the tubers. Most of these pests can be easily controlled by the use of traps or poison.

## 4.12 CEREALS

Cereals (e.g. wheat, maize, rice, millet, sorghum) belong to the grass family. They all have three things in common:

(1) They are consumed in all parts of the world in one form or another.
(2) They all produce inflorescence of male and female flowers on separate parts of the plant.
(3) The flour (milled grains) of all cereals can be easily mixed with one another for all types of purposes as human food (e.g. bread, cakes) and as livestock feed.

In West Africa the major cereals grown are sorghum, rice, maize, millet and wheat. The sorghums, millet and wheat will only thrive best above latitude 10°N and below latitude 10°S. The short season millet will thrive best at rainfall ranges too low for other cereals, while maize will thrive best at heavier rainfalls. Rice, on the other hand, can survive under conditions of heavy or light rainfall, on uplands or in swamps, as a sole crop or mixed with other crops on the same piece of land at the same time.

Rice (*Oryza sativa* L. – white rice; *O. glaberrima* Stend. – red rice)
Other names: Yoruba – *iresi*; Hausa – *shinkafa*; Twi (Ghana) – *emo/emu.*

Rice grows extensively in tropical to subtropical climatic conditions but will also thrive in temperate regions. It is used as food in different preparations; it is useful for starch production; and the stalk is useful as stock food, for bedding, for decorations and as fuel.

Rice thrives best in light to heavy acid soils, and unlike most plants it thrives in swamps because of the capacity of the roots to survive in soils with a high sulphur content. Modern breeding work has been successful in producing high-yielding varieties most suited to different ecological conditions, with yields under experimental conditions lying in the range 4·8–7·4 tonnes/ha (1·9–2·9 tons/acre) of dry grains in West Africa. Cultivated varieties such as *upland rice* (OS6 in Nigeria, Anethoda in Sierra Leone), *swamp rice* (Nachin 11 in Nigeria, CP4 in Sierra Leone) and *floating rice* (Agbedo for the Niger–Bane confluence in Nigeria) are being continually improved through the efforts of several agricultural research institutions in West Africa, such as the International Institute for Tropical Agriculture (IITA) in Ibadan, Nigeria and the West African Rice Development Association (WARDA) in Monrovia, Liberia.

*Land Preparation and Planting:* The land has to be carefully hoed or ploughed long before the rains. No ridging is required, but planting in ridge furrows when rice is planted as an intercrop is also possible. For the swamp variety, a nursery bed carefully raised, well aerated and made with humic topsoil is required.

Planting can be by broadcasting, especially for the upland variety, while the nursery bed is used first before final transplanting in the case of the swamp and floating varieties. Transplanting is usually done at the 4–5 leaves stage or at 15 cm (6 in) height in July–August in West Africa, depending on how early the rain starts. Planting is at 10 cm (4 in) intervals, in rows and usually in not more than 3 cm (1 in) of water.

Upland rice requires weeding at the early stage of growth and again immediately before flowering if considered necessary. The swamp variety requires no weeding, but obvious shrubs and persistent weeds should be cleared before transplanting.

Fertilisers (N–P–K and especially phosphatic fertilisers) considerably improve the yields of all varieties of rice (by up to 30 per cent). Application of fertiliser should be done in two or three stages for maximum yield, especially immediately before and during the flowering phases.

*Harvesting, Milling and Storage:* In 5–6 months after sowing or transplanting, rice is normally ready for harvesting. In Ghana maturity may take only 4 months or even less, and new varieties are being developed which mature earlier. The mature heads are cut with a machete or scythe, dried and kept in store. Average yield varies like that of any other crop, depending on several factors: (a) the variety planted; (b) the area or soil and ecology; (c) whether the rice is upland type or swamp type (swamp types yield relatively more than upland types); (d) whether cultural operations were adequate, timely and well carried out; and (e) the level or amount of pest and disease attack. On average, a yield of 0·7–1·7 tonnes/ha (0·3–0·7 tons/acre) of paddy is possible with upland rice, while swamp rice will yield 2–4 tonnes/ha (0·8–1·6 tons/acre).

After harvesting, rice can be stored easily as paddy provided the heads are well dried. If rice has to be milled, processing sometimes starts with *parboiling* in the traditional manner. After parboiling, which is a kind of steaming to loosen the fruit coat (called *husk*), the paddy is sun dried and then milled. The milled rice can be sold or cooked for home consumption. Wastes from milling are referred to as *rice bran* and usually fed to livestock or used as manure.

Parboiling rice is a delicate operation. If it is overdone, the rice becomes cooked and not easy to mill. If it is underdone, the husk is still more difficult to mill and most of the rice grain is broken. If drying after parboiling is slow, the rice becomes rotten and smelly – offensive odour! If drying is carried out on the ground or mats, care has to be taken to avoid the addition of impurities such as stones, animal wastes and weed seeds.

Milling also is a delicate operation. If it is poorly done, much of the rice may become broken and will be lost with the bran. Poorly ground rice fetches only a poor price in the market.

*Diseases and Pests:* Fungal diseases such as *rice blast* (*Pyricularia oryzae* Cav.), *brown leaf* and *flower mould* (*Helminthosporium oryzae* Breda de Haan), *green smut* (*Ustilaginoidea virens* Cooke), *leaf smut* (*Entyloma oryzae* H. and P. Syd) and *black spots of rice* (*Nigrospora oryzae* Berk and Br. Petch) all attack rice seedlings and mature plants, causing untold losses – low yields, poor harvests and bad financial returns.

*Rice blast* is one of the most important fungal diseases of rice. It attacks at the seedling stage. Attack is indicated by small brown spots which later elongate into grey spots with brown margins. Eventually, as attack worsens the seedling leaves dry up and the plant dies. On mature plants the attack is shown by discolorations on plant stalk and panicle. Few rice grains are produced, and these too become discoloured and pale. Control is by planting resistant varieties.

The pests that attack rice are *birds*, *bush rats*, other *rodents* and *moths*. Moth larvae bore into the stem causing damage to the stalk, and the plant dies. Some adult *bugs* also suck ripening heads of rice paddy. Birds (e.g. *Quelea* sp.) attack rice at the milling stage, sucking the juice and eating the grains. Other birds and rodents eat the rice seeds when planted, especially when rice is planted by broadcasting.

Some pests can be easily controlled by rotation and clean husbandry. Birds may be controlled by using varieties that mature early before the migratory birds arrive. With other pests, control is by trapping, killing or using scarecrows.

Maize (*Zea mays* Linn.)
Other names: Yoruba – *agbado*; Edo/Urhobo – *oka*; Ewe (Ghana) – *ebli*.

Maize, sometimes referred to as *corn* (American), is one of the most important staple food crops in West Africa. Because the crop requires a lot of rain, a minimum annual rainfall of 75 cm (36 in), it gives the best yield when cultivated in the tropical belt, extending as far north as the Guinea savanna (approximately 10°N) of West Africa.

Maize is also an important ingredient of livestock rations. The green plant may be cut before fruiting and used for *silage*. For food it is prepared in different forms. As a green crop the maize is roasted or boiled; when fully dried it can be cooked and used in the form of maize powder, maize meal and general confectionaries such as bread, cakes and biscuits.

For different uses, different types and varieties are cultivated over most parts of West Africa. The most common ones are *dent maize*, *flint maize*, *sweet corn*, flour or *'soft'maize* and *popcorn*.

*Land Preparation and Planting:* Maize is a tropical crop that needs a lot of rain at its early stages. Hence the soil preparation and ridging have to be completed before the rains start, in about March in most parts of West Africa. Maize can be grown on the flat with gently sloping farmland. However, if the slope is steep it is advisable to use hills or ridges, both to check erosion and to increase yield by increasing the topsoil area available to the individual plant.

The maize may be planted either as a sole crop or as part of a mixed cropping system where it is usually intercropped with okra or any other vegetable or interplanted with cassava. When it is intercropped with okra or any other vegetable, the latter crop is harvested before the maize. However, when it is interplanted with cassava, the maize is harvested first and the cassava later.

Maize is usually planted direct on the field at the rate of 3–4 seeds per hole at a depth of 2–4 cm (1–1·5 in) at a spacing of 30–38 cm (12–15 in) along the row, and on ridges 90 cm (3 ft) apart. Seedlings take 3–5 days to germinate. To ensure an adequate plant population, it is always advisable to carry out a germination trial.

Planting in West Africa is from early March to April. Thinning of plants to one per stand or substituting for missing stands is best carried out when seedlings are 2·5 cm (1 in) high. Weeding should be done after maize plants are fully established to avoid damage to the delicate roots. Two weedings should be sufficient before harvest: the first 3 weeks after germination, and another 4 weeks after the first.

Application of fertiliser, especially the N–P–K combination, is usually done in the hole round the plant. Response to fertiliser is high but varies according to variety, time of planting, time of application, rainfall and soil type. Therefore for maximum response it is always advisable to plant the variety suitable to the area, especially with respect to soil, rainfall and previous cropping. Phosphates are best applied at the time of sowing as a base dressing, but nitrogen should only be applied when plants are over 50 cm (20 in) high.

Male inflorescence appears as the terminal feature in maize when the plant is mature. Later the female flowers appear below, in the leaf axils between the stem and petioles. The plant is *hermaphrodite*, i.e. produces male and female flowers separately on the same plant. The *tassels* (male pollens) are shed before the female *silks* of the same plant are ready for fertilisation. Hence maize plants are naturally cross-pollinated.

In the wetter parts of the equatorial rain belt, where the rainy season is long enough (February–November), two crops of maize are produced each year. The early maize is in March/April–August, while the late maize is in September–December/January, depending on the rains. However, as a control measure against pests and diseases, it is not advisable to grow the two crops of maize on the same or adjacent piece of land in any one year. In most other areas, only one crop can be cultivated each year.

*Harvesting and Processing:* Maturity usually takes from 100 to 140 days depending on the variety, but there are also some 60-day varieties, for example in Ghana. However yields are generally lower because of the relatively low amount of water available during the period. The harvesting periods in West Africa, where two crops a year are possible, are August–September for early maize and December–January for late maize. Where only one crop per season is grown, harvesting is usually between September and October.

Yield varies depending on variety, weather, moisture content of maize at harvest, whether early or late crop and, of course, whether fertiliser was applied. Early maize can produce up to 2 tonnes/ha (0·8 tons/acre) of dry grain with a moisture content of about 20 per cent. The late crops can produce up to 0·9 tonnes/ha (0·4 tons/acre) of dry grain and usually with a lower moisture content. Average yield in areas capable of only one crop per season is 0·9–1·2 tonnes/ha (0·4–0·5 tons/acre) of dry grain. Where green maize is harvested, yields are difficult to calculate, because the fruits are harvested gradually as they become mature.

Once maize is harvested and dry, it is stored. Storage may be with or without the *sheath* (the outer leaf cover), or sometimes *shelled* if it is dry enough or even on the *cobs* (unshelled). Traditional storage methods include the following:

(1) Sheathed maize may be stacked on a raised platform in the field, in *rhumbus* (granaries) or in special constructions made for it.
(2) Maize on cobs can be stored in rhumbus or in bags, baskets or any other convenient containers.
(3) Shelled maize is usually stored in containers of different types ranging from simple raffia bags to metal drums. Seeds for future planting are shelled and stored in tight containers such as bottles, or left on cobs and stored in front of the fireplace.
(4) Modern storage involves the use of special constructions such as silos, mesh wire cages, storage bins and the like. These are expensive. Their advantages lie in that they are able to prevent pest attack and to keep the stored maize in good condition until it is needed.

Processing of maize into food for human consumption and animal feed is very important. The milling of maize into powder (maize meal) involves grinding, winnowing and fermenting, especially in the making of maize pap or porridge and other foods.

*Diseases and Pests:* Quite a number of diseases and pests attack maize in the field and reduce yields. A good number of pests also feed on harvested maize in storage and reduce the available maize that can be sold or used for food.

Diseases of maize that are of common occurrence in West Africa include *rust (Puccinia polysora),* a fungus disease producing red spores which are easily noticeable on the leaves. The fungus reduces the amount of green area of the leaves, thus reducing photosynthesis. This in turn reduces food manufacturing, and the yield if any is usually very poor. The only effective control is to plant resistant varieties. Uprooting and burning infected stands reduce the spread of the disease. Rotational cropping reduces the spread to other crops.

*Maize smut (Ustilago maydis)* is also common. Another fungus disease, it attacks the mature grains, leaves, stems and tassels of maize. In severe attacks, the affected parts are identified by 'boils' like protrusions or swellings (spores of the fungus) on the affected tissues. The only effective control is by uprooting and destroying infected stands. Using resistant varieties is an advantage.

*Maize streak*, though not very common, can be a very deadly epidemic on maize farms. It is a virus disease. Symptoms of infection include the presence of numerous short broken streaks on the plant. Growth is stunted and vegetation is poor, resulting in highly reduced yield. No effective control method is known. Burning or burying infected plants may help to reduce the spread of infection. Destroying the insect vector (*leaf-hoppers*) also helps, as does the practice of crop rotation.

A lot of pests attack maize at both seedling and adult stage, both in the field as well as in the store.

*Army worms* are the larvae of several species of moths. The moths lay their eggs between the sheath and stem of the maize. The larvae from such eggs feed on the growing points of the plants and the leaves and also burrow into the stem, causing both defoliation and a collapse of the stem. The plant usually dies if there is no control. Control is by hand picking the worms and by spraying with insecticide.

*Earworms* are larvae of moths also. In this case the eggs are laid on the female inflorescence (i.e. the silk), and the larvae eat out the developing grains, causing great losses.

*Rodents* and *weevils* eat maize grains both on the field and in the store. Weevils, rodents and also *beetles* can be controlled by timely harvest and adequate storage. Fumigation in the store with insecticide also reduces infestation.

Guinea corn (*Sorghum vulgare*; *S. guineense* Stapf)
Other names: Hausa – *dawa*; Yoruba – *oka-baba*.

Guinea corn is one of the most common cereals in West Africa, where it is eaten in one form or another by many more people than are the rest of the cereals put together. It can be roasted or ground and made into meal, like maize, which is a close diet alternative or substitute. *Millet*, especially bulrush millet (*Panicum cinereum* Stapf and Hubbard), is a very close substitute, except that millet is more resistant to drought than is guinea corn. Besides, the early-maturing type of millet requires only a short period of rain (sometimes 80–90 days) for full maturity.

Climatic suitability for guinea corn is mainly the Guinea and Sudano–Sahel savanna types. It will survive in areas of low rainfall, between 50 and 63·5 cm (19·5–25 in), where maize, a rain-loving cereal, will not thrive. In poorer soils, sorghum will yield relatively more than millet, and it can withstand drought owing to its very well-developed root system.

There are several varieties of sorghum, some of which are especially adapted to different localities. These different varieties show differences in growth period (i.e. from seedling to maturity), size of plant (height and tillering), compactness of the panicle, and grain colour and size.

Sorghum can be planted as sole crop or in a mixture with millet or rice.

*Land Preparation and Planting:* Land is usually prepared before the rains, which come usually between June and September in the savanna belt. Farmyard manure and fertilisers can be applied before or at planting and/or during the early stages of plant growth.

Ridges are usually advisable, more to check erosion than for anything else even though guinea corn and millet can be planted on the flat on gently sloping farmland.

Seeds are planted, usually in May, at 12–15 cm (4·5–6 in) intervals in rows usually 60 cm (2 ft) apart. Weeding is very important, especially at the early stages of growth and twice more later before flowering, after which weeding may no longer be advantageous, especially on farms with a good virile growth of the plant.

*Harvesting and Storage:* After 5 months, guinea corn is usually ready for harvesting between October and January. Yield varies according to variety and soil conditions but is generally 0·45–0·75 tonnes/ha (0·2–0·3 tons/acre) of dry grain. Yields of 0·8–1·2 tonnes/ha (0·3–0·5 tons/acre) of dry grain are possible.

To harvest, the top (head) of the plant, where the fruit is, can be cut with a machete or machines if the plants are in rows. The stalks are tough, straight and useful in making baskets, mats, fences, barn doors and beds and for other traditional uses. The dried leaves can be fed as hay or used as bedding for livestock.

After harvesting, the heads can be dried and stored in a granary (rhumbu). If used or sold immediately, guinea corn has to be first threshed in a mortar then winnowed, so that the chaff is separated from the grain in the traditional way. The grain can be pounded or powdered and used for food as required. The chaff as well as the grain is fed to livestock.

*Diseases and Pests:* A number of fungal diseases afflict guinea corn and millet. The more common ones in West Africa include *smut*, *mildew* and *rust*. A number of bacterial diseases (e.g. bacterial blight, leaf spot) also attack guinea corn and millet.

*Grain smut*, caused by *Sphaceolotheca sorghi* (Link), is found on many guinea corn farms. The affected heads have no grains; they are empty and light in colour, and instead of grains there are thick-walled greyish cylindrical spore sacs usually filled with black spores. Yield is low, and if infection is high the crop may not even be worth harvesting. Control is by the use of resistant varieties, clean husbandry (especially clean seeds) and the practice of crop rotation. Diseased heads should be cut off and buried before the spores are formed, to prevent spreading.

Other types of smuts include the *open smut* or *head smut*, in which part of the fruiting head is enclosed in a big spore sac filled with black spores.

*Bacterial blight* is easily recognised by yellowing of the leaves, internal discoloration of the stem and the sudden death of the plant. Leaf spots of different kinds (e.g. *leaf spot*, *eye spot*, *anthracnose*, *leaf blight*, *sooty stripe*) all affect guinea corn and millet, causing varying degrees of losses from very small to complete crop failure if left unchecked. Effective control again, is by the use of resistant varieties and the practice of crop rotation. Spraying may help if infection is detected early enough. As a golden rule, infected stands should always be uprooted and destroyed.

*Birds* are the most serious pests. *Rodents*, *weevils* and *stem-borers* attack cereals in general, and guinea corn and millet are no exceptions. Several species of *bugs* (the Hemiptera, e.g. *Dysdercus* spp.) that feed on cotton also attack the flower of sorghum, sucking the juice at the 'milky stage' and thus later forming empty heads.

*Stiga* are a group of parasitic plants that grow on the roots of guinea corn and feed on the growing plant. Control is by hand picking and burying the offending parasite.

## 4.13 LEGUMES

Family Leguminosae comprises a lot of useful plants that are cultivated for their seeds, or as green manure crops because of the nitrogen-fixing ability of their roots, or as cover crops or even as pastures for the balanced diet of herbivorous animals.

The more common food-crop legumes, sometimes called *pulses* in West Africa, are groundnuts (*Arachis hypogaea*), cowpeas (*Vigna unguiculata* and *V. sinensis*), soya beans (*Glycine maximus*), pigeon peas (*Cajanus cajan*) and benniseed. Cowpeas and groundnuts are the two most important legume crops in West Africa.

The more common green manure crops are mucuna, calapogonium and centrosema, which serve as cover crops when growing and are later ploughed in (see notes on the nitrogen cycle, section 3.15).

Legumes such as stylosanthes and centrosema arc very common as pasture mixtures since they improve the nitrogen content of the herbage for livestock feed.

### Cowpeas (*Vigna unguiculata* Walp)

Other names: black-eyed beans; Yoruba – *ewa*; Twi – *edua*; Nupe – *ezo*.

Cowpeas, an annual legume, is the most important, cheapest and most widely consumed legume crop in West Africa. This makes it the single most important source of cheap vegetable protein in West Africa.

Several varieties are cultivated, each variety being most suited to certain local areas. The *testa* (seed coat) may be white, brown or mottled in colour. Vine stems may be prostrate, spreading, climbing or dwarfy.

The most suitable climate is when annual rainfall is 75–150 cm (29–58·5 in), and the best soil is fertile loam. Ideally soil has to be just slightly loose. This explains the successful cultivation of cowpeas in the wetter northern areas of the Sahel savanna of West Africa. The crop is slightly drought resistant, and it becomes highly susceptible to a number of root diseases as well as pests in the wetter tropical climate in the southern parts of West Africa.

*Land Preparation and Planting:* Cowpea does not require any elaborate land preparation. It can be planted as a sole crop after the early maize crop in the rotation or as a mixed crop with guinea corn, cassava and yams.

Planting is done usually in April–May in the coastal areas with earlier rains and in July–August in the north. Two or three seeds are sown per hole depending on the germination percentage. Seeds are planted 25–30 cm (10–11·5 in) apart in rows which are themselves 75–92 cm (2·4–3 ft) apart. Germination occurs within 3–4 days and maturity is reached in 2–3½ months for the early-maturing variety and 3–5 months for the late-maturing type, depending on variety and rainfall.

Early weeding is advised until the crop covers and chokes out weeds.

*Harvesting and Storage:* Harvesting is done by hand, by picking the ripe pods just before they open naturally. There are usually several pickings before harvesting is complete. The harvested pods are stored in rhumbus or baskets. The ***haulm*** (the vegetative part) is fed to livestock as hay or fodder.

The yield varies, again depending on variety and soil conditions. Yields of 0·25–0·6 tonnes/ha (0·1–0·2 tons/acre) of dry grain (out of pod) are common in West Africa but a yield of 0·8 tonnes/ha (0·3 tons/acre) is possible.

Storage is very important. Because cowpeas are rich in protein, the crop attracts a lot of pests both on the field and in the store. When fully dried, the pods can be stored in baskets and kept in dry conditions to check mould, fungus and mildew. Dried pods are easily shelled either by hand or by threshing in the traditional way in a mortar. Shelling machines may also be used. While in store, fumigation with insecticide is of great advantage.

*Diseases and Pests: Mosaic,* a virus disease, gives rise to mottled and distorted leaves. Control is by roguing and destroying infected plants.

*Leaf spots* of several kinds, caused by *Cercospora* fungi species, are also common. Control is by the use of resistant varieties.

*Rust* (*Uromyces vignae* Barclay) has been found in a few cases but only doing little damage.

Pests constitute the source of greatest losses of cowpeas. Virtually all pests known – *birds, moths, butterflies, rodents, eelworms, beetles, bugs* – attack cowpeas. Moths, for instance, feed on the leaves. The adults of some beetles feed on the leaves and on the seeds both on the field and during storage and even right into the market or house. Butterfly larvae feed on the leaves and pods. Weevils, especially *bruchids* feed on the pods at maturity both on the field and in the store and even into households.

Control of pests is by spraying on the field and in the store with insecticide. The practice of crop rotation reduces the degree of infestation.

Groundnut (*Arachis hypogaea* Linn.)
Other names: peanuts; monkey nuts; Yoruba – *epa*; Twi – *nkate.*

A valuable export crop for most West African countries, groundnuts are easily the most important source of vegetable oils in West Africa. It is for this reason that groundnuts are widely cultivated in all parts of West Africa.

The crop is very adaptable, being cultivated between the latitudes 40°N and 40°S, thereby covering both tropical and sub-tropical climates. Two main varieties are grown: (a) the bushy upright type; and (b) the trailing spreading type which has its nuts (the fruits) buried in the soil in a random pattern. Both types do very well in rich, light sandy-loam soils. The annual rainfall requirement is 50–125 cm (19·5–49 in), well distributed throughout the growing period of the crop.

*Land Preparation and Planting:* Ridges or hills can be an advantage, especially for the upright variety. The spreading variety will do equally well on the flat or on ridges. Planting of two to three rows on each ridge is common. Spacing is 10 cm (4 in) between plants for the upright type and 15 cm (6 in) for the spreading type. Rows should be 60 cm (2 ft) apart.

Planting is done during the early rains, in March–April in the southern tropical and savanna areas and in May–June in the Sahel savanna area and beyond in West Africa. Groundnut can be planted as a sole crop or interplanted with yams or even cassava, with the cassava coming in later. The crop matures in 4–6 months. Weeding should be done regularly, two or three times before harvesting, especially for the upright variety. Two weedings at the early stages is usually enough for the spreading type as the good vegetative growth very soon covers out weeds.

*Harvesting and Storage:* Harvesting is done by first spiking the soil, to loosen the soil, ease the roots and make uprooting of plants very easy. Then the seeds are picked out by digging the soil around the roots. The vegetative part of the plant is good fodder for livestock.

The seeds are dried and stored with the shell. Sometimes shelling is done immediately, where the seeds have to be sold to government agencies or to the oil extraction factory. Shelling can be done by hand, by threshing in the traditional way, or by the use of shelling machines.

Yields of 0·4–0·6 tonnes/ha (0·16–0·24 tons/acre) of shelled nuts are considered average, with yields of 0·8–1 tonne/ha (0·3–0·4 tons/acre) possible. With better husbandry and variety, yields can be considerably increased.

Storage is very important because groundnut is a high-priced crop. In Nigeria, for instance, groundnuts purchased by the government through the marketing boards are stored in bags which are then stacked in the open into pyramid shapes. These are then covered with tarpaulin sheets and fumigated. When well treated in this manner, groundnuts can be stored for nearly a year. Storage in bags, baskets or pots is also possible, and if there are small quantities these can be stored in closed stores which also have to be fumigated or sprayed with insecticide.

*Diseases and Pests: Rosette*, a virus disease, is the most deadly disease of groundnuts. It can easily lead to a complete loss of harvest. This disease is carried by an aphid which sucks on the plant. Infection leads to stunted growth with very marked mosaic leaves. Chlorotic leaves with dark veins usually indicate the beginning of infection. Control is by use of resistant varieties, spraying with insecticide to kill the aphids and the practice of crop rotation.

*Seed rot*, a *Rhizopus* sp. fungal disease, is also common. It destroys the seed before germination.

*Aspergillus flavus* causes *crown rot*, which destroys the germinating seedling by killing the hypocotyl. The fungus lives inside the seed pods and produces toxic substances (*aflotoxins*) which are poisonous and can kill animals that may feed on the nuts. Control is by the use of fungicide chemicals or dusts.

*Cercospora* fungi also attack the leaves and cause dark brown spots.

*Wilt* is another fungal disease in the soil, which causes a die-back of seedlings. Control is by roguing and burning infected plants.

Pests on the field are not many since the fruit is usually buried, but in the store all types of *beetles* and *weevils* attack groundnuts. *Aphids*, as mentioned earlier, suck on the plant. The aphids are vectors of the virus rosette disease. *Millipedes* sometimes burrow in the soil to eat the nuts. By far the most important pests are weevils. They bore holes through the seeds, feeding on the cotyledons and often damaging the growing parts of the fruit. Adequate storage is called for.

Control of pests is by good drying of the seeds to reduce fungal attack. Spraying or fumigating in the store is essential for large storage. Traditionally, groundnuts are stored in tins and baskets that are well protected from these pests.

## 4.14 INDUSTRIAL CROPS

With the increase in commercial agriculture in West Africa, subsistence farming goes hand in hand with the cultivation of some crops exclusively for cash. Two of the more common ones are cotton, a fibre crop, and tobacco, a leaf crop. Both are annual crops which are being increasingly processed and used in local factories. Cotton is raw material for the cloth industry, while tobacco is processed locally into cigarettes, cigars, pipe tobacco and 'black-fat' (for snuff). These crops have the same ecological preference in that different varieties of both crops can be cultivated as virtual neighbours.

Of course there are a number of other industrial crops produced for both local and foreign use. Some (e.g. pineapple, ginger, sugar cane) are perennials, while others (e.g. oil palm, citrus, cocoa, rubber, cashew) are permanent crops and will be discussed later.

### Cotton (*Gossypium* spp. Linn.)

Other names: Yoruba – *owu*; Ivory Coast – *baoule gese*; Hausa – *abduga*; Twi (Ghana) – *asawa*.

*Gossypium*, a member of Family Malvaces, is indigenous to Africa, but the cultivated cottons of West Africa today are improved breeds from America.

Cotton has many uses as a fibre crop in threads, cloth making, ropes, absorbents etc. In industry it is used as yarn for weaving all kinds, colours and designs of clothing material. Because of its absorbent capacity, it can be dyed from its original white ('cotton white') into all shades of colour and can also be used as the wick in traditional household lamps as well as in modern fluid-filled lighters. It is one of the most important vegetable fibres.

In West Africa *G. hirsutum* is the most common variety, as compared to *G. barbadense* which is very common in the Nile Valley. *G. hirsutum* is a small plant about 1·5 m (5 ft) in height, with tap roots and several *bolls* or fruit (flowers). The leaves are broad, hairy and darker on the lower side away from the sun. It will thrive in any low rainfall area with or without irrigation, but it usually gives better yields with irrigation.

*Land Preparation and Planting:* Cotton can be grown as an intercrop with cereals or yams, but best yields are obtained when it is cultivated as a sole crop. It can be planted on freshly cultivated land on the flat or on ridges depending on the slope of the land. Ridges are useful for effective control of soil erosion and weeds.

Planting usually starts by May in the south and July in the far north of West Africa. Usually, 5–6 seeds per hole are sown, at a spacing of 0·9 m (3 ft) on ridges in rows that are 1·5 m (5 ft) apart. After germination, it may be necessary to thin or resupply missing stands so that there are only two plants growing at a spot. This gives the plants the best opportunity for good vegetative growth and encourages the yield of cotton.

Weeding is essential at the growing stage, and two weedings are recommended prior to flowering.

*Harvesting and Processing:* Cotton matures in 5–6 months after planting. Harvesting of bolls is usually done by hand in West Africa. It is possible to pick 10–18 kg (22–40 lb) a day. The boll is detached from the calyx and dispatched for ginning. Yields vary between 0·6 and 1·2 tonnes/ha (0·2–0·5 tons/acre) of seed cotton, depending on the variety planted and whether it was planted as sole crop or in a mixture with other crops. Yields of 2 tonnes/ha (0·8 tons/acre) have been recorded.

Harvested crop is kept in baskets on the field, but when transferred to the store the crop is usually stored in jute bags.

Ginning may be done immediately or later. The ginning process, by which the seed and fuzz are separated from the more useful part (the *lint*), can be done by hand or machine. Special care has to be taken in ginning to avoid any damage or colour taint on the lint as this depresses the value of the lint and hence the price the farmer receives.

Cotton seed cake is useful cattle feed, and cotton seed oil is a useful disinfectant.

*Diseases and Pests: Bacterial blight* or *angular leaf spot* is the major bacterial disease of cotton and attacks at both the seedling and adult stages. When the seedlings are attacked, they become dry and brown and may eventually die. Infection of adult plants is shown in the leaves by small dark-brown angular leaf spots; petioles and stems show black sunken areas, and stems may eventually collapse. Bolls are also affected, and the seeds from infected bolls carry the bacterial infection over the next planting. Control is by quick sterilisation of cotton seeds before planting and using resistant varieties. Spraying may help, but roguing and destroying infected stands reduce the spread of the disease.

*Leaf curl* is a virus disease also known to affect cotton. The leaf curl virus, like most virus diseases in West Africa, is transmitted by *white fly*. Infected plants have stunted growth, thickened leaf veins and distorted young leaves. Control is by the use of resistant varieties. It is always advisable to avoid having the okra plant (*Hibiscus esculentus*) on the same rotation with cotton, as this virus disease attacks them both.

*Wilt*, caused by the fungus *Fusaricum oxysporum*, is easily transmitted in the soil and is often carried on the cotton seeds. Control is by soil sterilisation, which may be very expensive, and by the use of resistant varieties.

*Boll disease* is a fungus disease in which the boll affected is hard and the lint is stained. The disease is associated with cotton stainer bugs. When young bolls are attacked, they are shed, which causes great financial losses.

*Cotton stainer bugs* (*Dysdercus* spp.) are the most dangerous of cotton pests. The adult and nymph feed on the boll by piercing through them. This physical damage subsequently allows for the entry of fungi which attack, stain and damage the lint.

*Bollworms* are moth larvae which intermittently feed on the shoots and flower buds as well as the bolls, within which they tunnel and whose contents they eat out. Most of these worms also attack maize and okra, which are secondary hosts.

Control of pests is by hand picking, spraying with insecticide, and uprooting harvested cotton at the end of the 'closing' season, which in West Africa is usually by the end of February in the year following planting.

### Tobacco (*Nicotiana tabacum* L.)

Other names: Yoruba – *tâbâ*; Igbo – *anauni*; Hausa – *taba*.

Tobacco is a major industrial crop and is fast becoming a major source of cash income for some peasant farmers in West Africa. The leaves are harvested and processed into cigars, cigarettes, pipe tobacco and snuff. The nicotine concentration in the plant is high, so that the burnt ash from the plant is often used as an insecticide.

Several varieties are cultivated to suit different soil and ecological areas. The *Virginian hybrid*, for instance, prefers a light sandy loam soil and short period of rainfall. The leaves are harvested green and sun-dried or air-cured. Other varieties (e.g. *barley*) are cured as 'black-fat' for snuff, and other Virginian varieties are cured by fire as 'flue-cured' and used in different cigarette and pipe smoking mixtures.

In West Africa, tobacco is usually cultivated as a sole crop on small scale farms of 0·5–1 ha (1·2–2·5 acres) each, by a large group of farmers scattered over all parts of West Africa with Nigeria, Ghana and Ivory Coast as the leading producers. A few large farms do cultivate tobacco, but their relative total output is small.

Tobacco is an annual shrub-like plant growing to a height of 1·8–3 m (6–10 ft). Its importance as a cash crop lies in the broad leaves, which are harvested green. For the best performance and good leaf yield, tobacco is best sown on light well-drained soil, where there are high temperatures and a moderate (savanna) rainfall spread evenly throughout the year. In addition, tobacco requires calcium, potassium and organic manure as essential minerals for best-quality leaves.

*Land Preparation and Planting:* Tobacco is a delicate crop. Therefore, its requirements for ideal growth and yield are exacting and detailed. To start with a special nursery bed is essential. This should be prepared early in January–February from fine soil, preferably sterilised for soil nematodes and other soil-borne diseases and pests. Planting is done in drill rows 15 cm (6 in) apart or broadcast on the nursery bed which is usually 23 cm (9 in) above ground level and 1·2 m (4 ft) wide. The nursery bed can be any convenient length.

Immediately after sowing the nursery bed should be shaded with palm leaves. The shade should be gradually reduced until the last day before transplanting is done, when the nursery bed should be fully exposed to sunlight. This is done to give the young seedlings shelter from intense direct sun heat and from wind. Watering is best done twice a day: early in the morning and late in the evening.

Transplanting to the field is at the start of the rains, in April–May in the southern parts of West Africa but later (June–July) in the northern parts. Field planting is on ridges that are 1 m (39 in) apart and at spacings of 50–80 cm (20–31 in) along the ridge depending on the broadness of the leaves. Tobacco can also be planted at spacing of 60×80 or 90 cm (23×31 or 35 in) on the flat, provided the ground is reasonably level or only gradually undulating and not likely to cause erosion.

Transplanting is best done in the early morning, and water should be adequately supplied immediately after transplanting. N–P–K fertiliser combination should be applied when plants are over 2 weeks old in the field.

*Harvesting and Curing:* Tobacco leaves are harvested gradually as soon as the leaves mature and usually from the bottom up. Approximately 40 days after transplanting, the lower leaves are ready. The harvested leaves should be green and free from blemishes (e.g. cuts, bruises) and should be put on strings immediately for subsequent curing.

Yields vary depending on variety, averaging 0·5–0·8 tonnes/ha (0·2–0·35 tons/acre) of green leaves. To improve yield it is always advisable to practise 'topping', which is removing the top inflorescence as soon as this appears on most of the plants. After most of the leaves have been harvested, the remaining twiggy plant is uprooted and burnt. The ash can be used as disinfectant or insecticide or is burnt as manure.

*Curing* is a special operation by which the harvested green leaves are dried and preserved. Three curing methods are popular in West Africa. *Flue-curing* is done in specially constructed barns, and the hot air being circulated in the barn through flue pipes can be generated by the use of firewood, kerosene or even gas fires. With *air-curing* the process is simpler, using the natural atmospheric hot air. The leaves are hung on strings, which are then hooked to poles in the sheds until the leaves are dried but not crispy. *Steam-curing*, on the other hand, is specially used for curing 'black-fat' tobacco leaves which are used extensively in snuffs and other tobacco stimulants. Leaves are placed in a chamber and steam under pressure is forced through the leaves. Later the leaves are cooled and packed for use or sale.

*Diseases and Pests: Tobacco mosaic* or *leaf curl* is a virus disease which is perhaps the most deadly disease of tobacco. As with most mosaics, the vector is the white fly. Damage in this case is total loss since the affected leaf is not worth harvesting. Infected stands should be uprooted and burnt. Control of the disease is only through the use of resistant varieties, and plants should be uprooted immediately after the end of the tobacco season, which in West Africa is November–December. Where the incidence of leaf curl is high, tobacco cultivation should be discouraged. Failure to do so can only lead to an epidemic which may easily affect other crops.

'*Frog eye*' (*Cercospora nicotianae*) is another common disease of tobacco in West Africa. It is a fungus disease which damages the leaves and sometimes reduces the quality of the leaf. Extensive infection can lead to total crop loss. Attack is indicated by the presence of circular onion-margined spots having greyish centres (hence the name 'frog eye'). Control is difficult, but the use of fungicide (e.g. copper mixture) is usually recommended at the nursery stage.

*Black shank* (*Pythium aphanidermatum* Edson) is a fungus which attacks the collar of tobacco plants at ground level, leading mainly to blackening of the plant tissues. Control is difficult. Soil sterilisation helps, especially for the seed bed. The use of fungicide and resistant varieties is recommended.

*Root-knot* (*Meloidogyne* spp.) is an *eelworm* in the soil which attacks tomatoes, tobacco and a host of other crops. Hence it is important and desirable to undertake rigid crop rotation which will not permit the continued survival of the disease. Root-knot can lead to the death of plants and, if unchecked, to a total loss of crop harvest. Effective control methods include soil sterilisation, crop rotation and the use of resistant varieties.

## 4.15 VEGETABLES

With increased scientific knowledge, the study of farm crops and plants such as vegetables, fruits (orchards) and ornamental plants has become much more elaborate. A small book of this type cannot cover even the marginal areas of the subject, which is generally now referred to as *horticulture*. Nevertheless it is important to look at two of the most important vegetables in West Africa – okra and tomatoes – as well as to make a general note on the essential cultural practices in raising vegetables.

Vegetables cultivated extensively in West Africa can be grouped into two broad groups based on their origin. Vegetables of tropical origin (e.g. okras, egg plants, tomatoes, onions) are regarded as *local vege-*

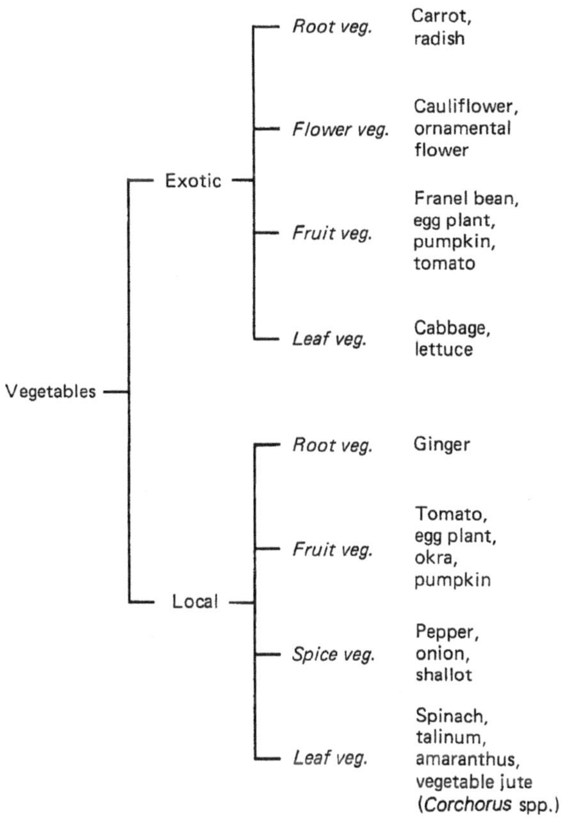


Figure 4.1 *Classification of vegetables.*

*tables*, while those originating from the temperate and semi-temperate zones but now being cultivated in several parts of West Africa (e.g. lettuces, cabbages, carrots, cauliflowers) are classified as *exotic vegetables*. Further detailed classification is possible, as shown in the Figure 4.1.

Most of the local vegetables, given the right care and moisture, will grow virtually anywhere in West Africa, but tomatoes of different varieties are best cultivated only under the most suitable soil conditions. The exotic varieties, on the other hand, require climates as close to the semi-temperate as possible. Hence most exotic vegetables grow much better at higher altitudes and better still in areas with short intense 'summer' heat and heavy rainfalls.

*Cultural Practices for Vegetables*

These cultural practices in general apply to almost all vegetables, whether local or exotic. The discussion will cover the areas of nursery operations, field (or vegetable bed) operations, planting, harvesting and marketing.

*Nursery Operations:* Most of the local vegetables can be raised from seeds or cuttings which can be planted directly on the field. Ideally, the best conditions are given when these vegetables are initially raised in the nursery and later put on the vegetable beds.

For some crops (e.g. tomato, lettuce) the use of a nursery is compulsory, partly because direct field planting may lead to a lot of seed waste and more so because these crops need special attention, especially with regard to moisture, heat, disease and pest control and cultivation, all of which can best be controlled under nursery conditions.

Nursery beds are small beds or pots made from humus. They have to be free draining. Sometimes soil sterilisation is done. Adequate water must be supplied regularly, and the seeds are usually planted in rows or broadcasted. The bed (or pot) is kept away from direct heat and sunlight, strong wind and pests. The nursery should be such that all seedlings have adequate space for maximum growth potential and are exposed to sufficient light. Soil around the seedlings should be pressed firm with the hand, so that seedlings grow up with strong stem, healthy roots and good foliage. Water should be supplied twice daily in sufficient quantities early in the morning and late in the evening. Weak or unhealthy-looking plants should be thinned out, preferably at the 3–4 leaf stage.

*Field or Vegetable Bed Operations:* Vegetable beds are best situated near natural water (e.g. river, spring, well) or where an adequate supply of tap or well water is assured. The site should be only very slightly sloping, with free-draining soil and an adequate amount of humus.

The beds, 1·5 m (4 ft) wide and of convenient length, are usually arranged against the slope and should be raised at least 15 cm (6 in) above ground level.

In preparing the bed, ideally the method of *double digging* is advised. This involves first carefully removing the topsoil to expose the subsoil. The subsoil is subsequently dug for more aeration and mixed with manure (compost or farmyard manure, FYM). All root stumps, debris and rocks are removed, and the topsoil cover is put back into position. This process is presented diagrammatically in Figure 4.2. This process of double digging should be done at least once a year. After double digging, the bed will be ready for planting only after 2–3 weeks, depending on the nature and stage of decomposition of the manure added to the subsoil. Additional fertiliser application to the top is advisable, depending on soil deficiencies and the type of vegetable to be planted.

*Planting and Transplanting:* Seedlings should be transplanted from the nursery to the vegetable bed before sunrise. Seedlings are dug out of the nursery with a ball of earth to keep the roots undisturbed and the collar intact. Seedlings are then replanted in the holes on the vegetable bed, firmly pressed in, watered, and if necessary shaded, but only very lightly. Shades should be gradually removed after 1 week.

Direct planting on the vegetable bed is best done in rows at the recommended spacing, depending on the type of vegetable. After germination, seedlings should be thinned. Daily cultivation between rows using a hand fork should be well done to aerate the soil and to permit drainage, to the advantage of the plant roots. Water should be applied after each cultivation. Other important operations include staking (e.g. beans, tomatoes), removal of infected leaves, and application of chemicals if and when desirable. In the case of citrus fruits raised on nurseries, operations such as budding, grafting etc. are also done while plants are in the nursery beds.

*Harvesting and Marketing:* Most vegetables are ready for harvesting within 2–3 months of planting. In most cases also, harvesting has to be done gradually as the leaves or fruits become ready. Delay in harvesting should be avoided, since over-ripe vegetables do not store; their quality deteriorates rapidly, and consequently they are sold at lower prices in the market. Leaves, roots and fruits, depending on the type of vegetable, should be harvested early in the morning and taken to the market for sale. However, if the market site is far away and transportation is required, it is advisable to harvest the previous evening and allow the fruits, leaves or roots to stay out in the open overnight in good containers (e.g. baskets, buckets).

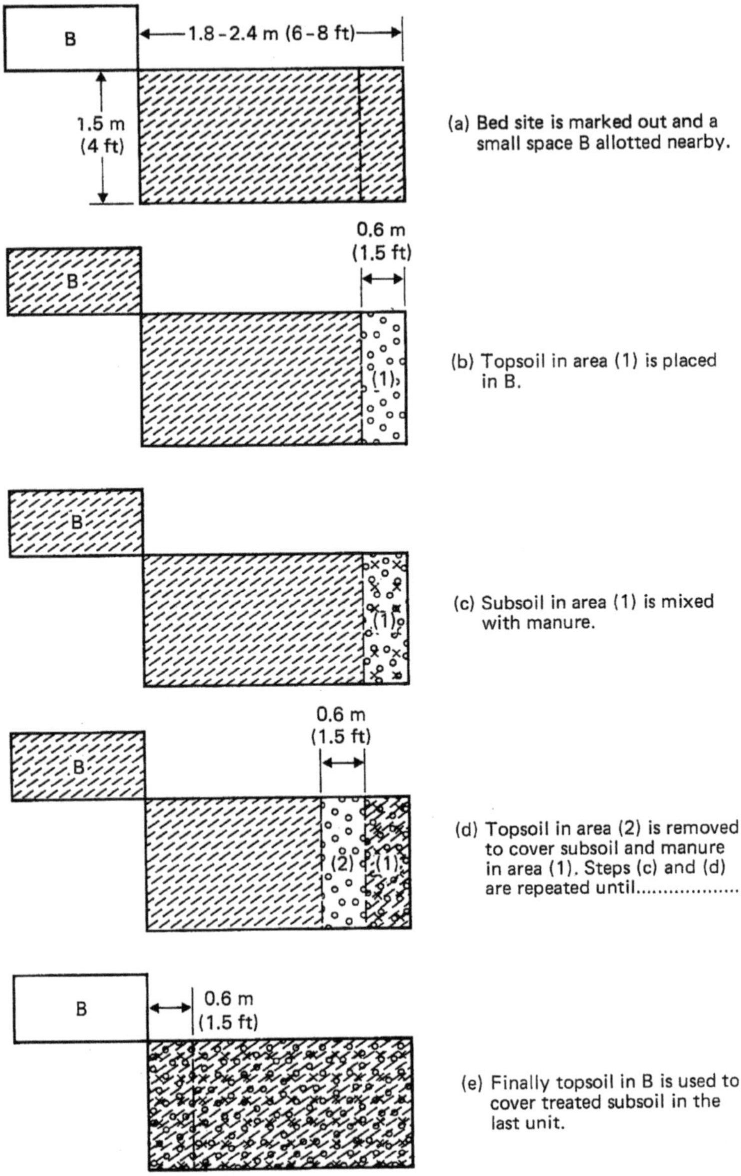


Figure 4.2 *Preparation of vegetable bed by double digging.*

Bruising or tearing during harvest should be avoided as this will lower the market value of the harvest as well as render the harvest susceptible to infection. Harvested farm products should be cleaned, packaged properly and kept in good condition for the market. It is important to get to the market as early as possible, to keep the products fresh and thus to command better prices. While in the market, products should be kept away from direct heat or sunlight as this will damage their freshness and reduce their market price. Whatever is not sold should be carefully packed again and left in the open fresh air but away from direct sunlight or heat.

Damaged, rotten or decayed products should be discarded and made into a compost. If not too badly damaged, they are better used at home than taken to market. Nevertheless, certain customers may specifically request partially damaged products because they are cheaper. To meet such requests, such slightly damaged products should be packed and displayed separately from the good ones to avoid contamination or the poor prices for all crops that may happen if good and bad crops are mixed, packed, or displayed together.

### Tomato (*Lycopersicum esculentum*)

Tomato, a member of Family Solanacea, is cultivated for its fruits which when ripe have a golden yellow or blood red colour depending on the variety. Tomato fruits can be eaten fresh, sliced and used in the preparation of salads, milled and made into pastes and added to stews, or crushed for their juice. Tomato fruits are rich in minerals and vitamins.

*Land Preparation and Planting:* Tomato seeds are better raised in nursery beds and later transplanted on to vegetable beds. High yields are possible when small quantities of nitrogen and higher amounts of potash are used as the plants become mature. Transplanting is best done at the 4–6 leaves stage when plants are 10–15 cm (4–6 in) high. Spacing, depending on variety, is 40–45 cm (15·5–17·5 in) apart on rows 1–5 m (3·5–16·5 ft) apart. Staking should be done early before flowering, to train the stems erect as well as to avoid lodging when fruiting starts. Pruning is recommended as the flowers set seeds, to avoid too many small fruits. Experience is always the best guide on this matter.

*Harvesting and storage:* Harvesting should be at the start of the ripening of fruits. Harvested fruits should be disposed of when fully ripe and clearly not more than a week after harvesting, to avoid rotting and falling prices for poorer quality products.

*Diseases and pests:* A number of virus diseases attack tomatoes. *Root-knot*, for instance, affects tomato plant roots, causing a number of swellings (*galls*) at root tips and leading to poor yield and total loss of crop. Control is achieved by rotational cropping on the vegetable beds, the use of resistant varieties and, where funds are available, soil sterilisation.

A number of fungal diseases plague tomato, affecting leaves, stems and often the fruits. The more common ones are *leaf mould* (*Cladosporium fulvum* Cooke), *leaf spot* (*Septoria lycompersici* Speg.) and *fruit rot* (*Selevotium rolfsii*). Control is by the use of fungicide and clean husbandry.

Some pests also attack tomato. Important ones are the nymphs and adults of several Hemiptera *bugs*, which feed on the leaves and suck on the fruits, and the larvae of several *moths* and *butterflies*, which also feed on the leaves and fruits. Control is readily achieved by hand picking the pests, but if they are in large quantities spraying with insecticide may help.

Okra (*Hibiscus esculentus* L.)

This vegetable will grow in any part of West Africa. However, different areas have special preferences as regards the viscosity of this fruit, while in others okra is cultivated mainly for factory processing. Two fruit types cultivated extensively in West Africa are the short rounded-base type with a tapering end and the long finger-like type.

*Land Preparation and Planting:* Seeds can be sown directly on the field at the beginning of the rains as sole or mixed crop or directly on the vegetable beds. Three seeds per hole are usually sown, 0·6–0·9 m (2–3 ft) apart in two rows per vegetable bed. Thinning is done 1 week after germination.

*Harvesting:* Maturity occurs 2–3 months after germination, and it is advisable to harvest fruits gradually as they become mature, to avoid the toughening and hardening of fruits.

*Diseases and Pests:* Two important virus diseases of okra are *leaf curl* and *leaf mosaic*, whose attack gives rise to stunted growth and leaf emaciation. Control is by the use of resistant varieties, the practice of crop rotation (okra not to follow cotton and vice-versa) and the use of insecticide to kill the *white fly* vector.

A few fungal diseases such as *powdery mildew* (*Oidium* sp.) and *black leaf mould* (*Iremopsis aciculosa* (Wint.) Stew) attack both the leaves and fruits of okra. Control is through clean husbandry, the practice of crop rotation and the use of fungicide.

## 4.2 PERENNIAL AND PERMANENT CROPS

The major categories of perennials and permanent crops of importance in West Africa include: (a) *oil plants* (e.g. oil palm, coconut, shea-butter tree); (b) *plants for beverages, spices and drugs* (e.g. cocoa, coffee, kola, date palm, locust bean, pepper, ginger, sugar cane); (c) *fibre plants* (e.g. kapok, sisal, silk cotton tree); (d) *latex trees* (e.g. rubber); and (e) *edible fruits and nuts* (e.g. citrus, banana, mango, pawpaw, avocado pear, cashew, pineapple).

### 4.21 OIL PLANTS

Oil Palm (*Elaeis guineensis* Jacq.)
An indigenous forest product of West Africa, the oil palm is ubiquitous as a natural grove, and quite recently commercial plantations have been established with improved and hybrid varieties. The oil palm belongs to Family Palmae and in a natural grove attains a height of 15–18 m (50–60 ft). The trunk is formed, as with most monocotyledons, by spirally arranged leaf bases from which the leaves soon detach as these wither away. The leaves, consisting of many (nearly 100) leaflets, are large and usually 1·5–1·8 m (5–6 ft) wide at the widest point and 3·4–4·5 m (11–15 ft) long.

Nearly all parts of the plant have some traditional use. The root is used medicinally, while the trunk is good for house buildings, bridge construction and fuel. The leaves are used for roofing (thatching), for brooms (from the leaflets' sub-laminate) and, in the young (yellow) immature state, for ceremonial branches on Palm Sundays and at traditional worships along the coastal parts of West Africa. The inflorescence is tapped for palm wine, while the fruit if allowed to ripen is a useful supply of palm oil, palm kernel and other subsidiary products.

Rainfall is the most limiting factor to the extensive cultivation of the oil palm. It requires about 150–75 cm (58·5–68 in) of rainfall a year, preferably evenly distributed throughout the year. Soil is also important. Best yields are obtained on soils that are slightly acidic and free draining and that have relatively high potassium and magnesium levels, with adequate amounts of nitrogen, chlorine, boron and phosphorus as well.

The natural groves usually have palms that bear ordinary fruits. Some bear red fruits with a brown cap; others bear red fruits with a black cap, while a third type bears green fruits. In all these types the *dura* (hard kernel shell) and *tenera* (thin kernel shell) varieties emerge, but these two extreme varieties are usually in small proportions and of poorer quality.

*Land Preparation and Planting:* Most of the oil palms now existing in West Africa are in the natural groves. However, the planting of improved varieties of oil palm in West Africa has been going on for the past two or three decades. Oil palm in plantations is now a common feature in all parts of West Africa.

The land required has to be freely drained and clear of undergrowth. It is common to have oil palm interplanted with maize, cassava and some local vegetables.

Seeds are planted in special nurseries call *germinators*. The seeds are planted in polythene bags and are artificially heated in these germinators for 80–90 days at temperatures ranging between 37·7 and 40·5 °C (100 and 105 °F). The germinated seedlings are then raised on nursery beds, under shade, with the start of the rains sometime in April–May.

The dry heat method of germination is faster and sometimes preferred. By this method, the seeds are cleaned, then soaked in clean water for about 7 days, dried in the shade for 1 day, and then packed in polythene bags which are then kept in germinators for 80 days at 29·4 °C (85 °F). This is followed by another 7 days of soaking and drying. After 2 hours of drying, the seeds are put in trays and watered lightly. This method ensures germination after 20–25 days.

Seedlings stay in the nursery for about another 12–15 months before they are transplanted on to the field.

Transplanting is done in the early morning or late evening in holes marked out at a spacing of 8–9 m (26–9 ft) in a triangular pattern. The holes are 1·2×1·2×1·2 m (4×4×4 ft) in capacity and partly filled with humic soil, and after transplanting the seedlings are firmly pressed in. Fertiliser may be added depending on soil needs.

Cover crops or intercrops of maize, cassava or legumes may be planted, and these are harvested before the oil palms build up a thick canopy. Cover crops and grass pastures are then usually planted, and these may be grazed by livestock provided enough care is taken not to damage the oil palm.

During the early days of growth, only ring weedings about 2 m (6·5 ft) in diameter are done. This is necessary to prevent choking the young plants. Legume leaves or pastures have to be regularly slashed or grazed. As the trees mature, regular pruning of leaves is essential. Footpaths should also be maintained for periodic checking and later harvesting.

*Harvesting and Processing:* Harvesting of improved varieties starts at the age of 4 years. Ripe fruits, usually at the sheath base of the fertilised inflorescence, have a distinct colour depending on the variety. The more common varieties produce blood red fruits capped with dark brown or black portions at their top end. Harvesting is done using a machete while the harvester is standing on a ladder, especially when the oil palm tree is too tall to be reached from the ground. Yields average 5–12 tonnes/ha (2–4·8 tons/acre) of fruit bunches.

Processing is either by the native canoe method or by the use of oil processors, which vary in their degree of sophistication and extraction rate. They range from the simple hand press handling 2·04–3·06 tonnes (2–3 tons) of fruit a day to the Stock Oil Mill press handling 204–408 tonnes (200–400 tons) of fruit a day.

The canoe method gives an extraction rate of only 60–5 per cent, while with the more sophisticated methods extraction rates as high as 85–90 per cent or slightly more have been reported. However, the native method allows for the production of fibre, which is good fuel.

After the oil is extracted, the kernel is washed and dried for 1 month or more by the traditional method before being cracked. The kernel seeds are sold to the marketing boards for export or for use in domestic factories. In oil mills the kernels are ready in a day or two. Kernels are useful as fuel too.

*Diseases and Pests:* Most of the diseases of oil palm, apart from mineral deficiency diseases, are caused by different fungi. The more important ones are collar rot, anthracnose and leaf spot.

*Collar rot* (*Ganoderma lucidum*) affects the collar at ground level causing rotting, and the plant wilts and finally dies. The fungi are visible if the tissues are exposed in the dead plant. There is no known control, but dead stands can be easily replaced.

*Anthracnose* attacks at the nursery stage. The seedlings show dark brown blotches on their leaves, and these dry out. If the disease is not controlled the seedling dies. Several fungi (e.g. *Botrydiplodia* sp. or *Carcospae elaeidis*) cause the disease. Control is by spraying the seedlings with fungicide.

*Freckle spot* or *leaf spot* (*Cercosporae elaeidis*) attacks the nursery and adult leaves, causing dark-brown leaf spots sunken in the centre and surrounded by an orange coloured hole. Leaves die and fall off. Without control the plant may die, particularly if attacked at nursery stage.

Some bacterial diseases (e.g. *seedling wilt, fruit rot, vascular wilt*) are also common.

Pests attacking the oil palm include the *oil palm miner*, which is a beetle whose larvae bore into the leaves and when in large numbers destroy the attacked leaves. *Weevils* also feed on the growing shoots, causing die-back of seedlings. Some Coleoptera adults also bore holes in the leaf rachis causing the death of the leaves, while some Hemiptera *fly* feeding on the fruits give rise to scale on the fruits.

Shea-butter (*Butyrospermum parkii* Kutschy)
The similarity between the oil palm and shea-butter tree lies in their usefulness for the production of fats and oils in all traditional forms. Similarly, whereas oil palm cultivation and natural ecology is confined to the forest groves of West Africa, shea-butter in its natural ecology is confined to the savanna belt of West Africa. By virtue of their importance, both the oil palm and shea-butter in their respective areas belong to the group of economic trees that are usually left uncut or untouched during the process of forest clearing.

Shea-butter oil from the seeds is used extensively as fuel for traditional lamps and as a solvent base for all types of cosmetics. The nuts are of wide domestic use. Shea-butter is an important cash crop in the savanna belt area of West Africa, even though its export potential is not proven yet.

*Cultivation:* The shea-butter is not cultivated either as mixed or sole crop in any type of plantation in West Africa. Rather, it is allowed to grow in its natural vegetation, particularly on fallow lands.

Shea-butter is not a very tall tree, barely reaching 6 m (20 ft) in height, but on compound farms it can reach a height of 15 m (50 ft), thereby confirming that with better care the tree can attain good growth and height.

*Harvesting and Processing:* Fruit bearing starts when the tree is over 16 years old with the fruits being allowed to mature fully and drop from the trees. The fruits are then collected, usually by women and children.

Collected fruits are first boiled to facilitate the extraction of seeds which are subsequently sun-dried for 6–10 days and then sold as shea-butter seeds or pounded. Pounding helps to remove the shell from the kernel. The kernel is then roasted and ground to form a whitish yellow paste. This fatty paste is further refined by using boiling water to leave a clear paste of shea-butter (Yoruba – *ori-amo*).

*Diseases and Pests: Leaf spot* (*Phyllosticta* spp.) is about the only commonly known disease and is a fungal attack. Symptoms include a series of rounded zonate spots on the leaves, especially the older ones.

A number of pests attack the shoot, leaves and fruits of the shea-butter tree. A number of Coleoptera *beetles* bore into the growing shoots and lay their eggs within. This may be extensive enough to damage the tree and to reduce its yield or kill it.

Furthermore, the *larvae* of *Civina butyrospermii* (Family Lepidoptera) feed on the leaves and, if extensive enough, can destroy all the leaves on the tree. Similarly, the larvae of other pests of the Lepidoptera and Pyralidae families feed on germinating fruits and pupate in the kernels, thus damaging the seeds and make them unfit for any use.

## 4.22 PLANTS FOR BEVERAGES, SPICES AND DRUGS

### Cocoa (*Theobroma cacao* Linn.)

Though not a native of Africa, cocoa is easily now the most important cash crop in West Africa. It probably originated from the valleys of the Amazon where it still grows as part of the natural vegetation. Ghana, for instance, is the world's biggest exporter of cocoa, where it contributes over 60 per cent of foreign exchange earnings. In addition cocoa is the chief export earner and the most important economic tree in the Cameroons, Nigeria, Benin (Dahomey), Ivory Coast, Togo and Equatorial Guinea (Fernado po).

Cocoa and kola belong to the botanical Family Sterculiaceae. Both serve as beverages and stimulants. Cocoa is used in the preparation of several industrial products: (a) food, as chocolate and cocoa confectionery; (b) drinks, as cocoa powder, cocoa milk and cocoa liquor; (c) cosmetics, in perfumes etc.; (d) soap making, from the husk ash; (e) livestock feed, as cocoa cakes; and (f) medicine, from the bark and roots.

In view of its importance in all countries of West Africa where cocoa can grow, the governments of these countries of West Africa have spared no funds and expertise in undertaking breeding work, agronomic studies, and disease and pest control for this crop.

Cocoa requires the best soil, well drained, with a good depth of topsoil which is porous and loose, has a low-based water table and is rich in humus. It thrives best in an equatorial tropical climate, with an annual rainfall of 100–140 cm (39–55 in), a temperature of not less than 17 °C (62 °F) and a relatively high humidity. It cannot withstand a long period of drought.

The cocoa tree is 8–10 m (26–33 ft) high at maturity and sprouts a good foliage canopy when cultivated on a plantation. A very good cocoa farm will provide an adequate amount of soil cover with its dead leaves and a foliage reasonable enough to inhibit undergrowth.

The important varieties cultivated in West Africa include the *West African Amelonado* (sometimes called *Forasthro*), *Upper Amazon*, *Criollo* and *Trinitano* varieties and several hybrids.

*Land Preparation and Planting:* Cocoa may be propagated either by seeds or vegetatively through cuttings. Land has to be cleared, and nursery shade and field cover have to be given enough time to establish before the seeds are planted. Nowadays sheds are built to serve as cover while the seeds, sown in plastic bags, continue to germinate, thus eliminating the need for nursery shade. It is also argued that closer planting on the field will eliminate the need for field shade crop. Both methods of field preparation are extensively used.

Where a shade crop is used, it is not unusual after clearing the land to plant plantain suckers approximately 3×3 m (10×10 ft) apart. Other foodstuffs (e.g. cassava, cocoyam, pepper) may also be incorporated provided these are removed before final transplanting of the cocoa seedlings. In other cases, cocoa beans are planted straight on the farm, usually at a triangular spacing of 15–20 cm (6–8 in). Three seeds are planted per hole, and thinning is done later.

Alternatively, the 'bare root' seedlings method involves first raising the seed in the nursery, using plastic bags or baskets or nursery beds. Cocoa beans are planted close together on the beds or at the rate of one bean seed per bag. The nursery bed soil or soil in the bag has to be rich in humus and free draining.

After 2–3 months or more, not exceeding 6 months, the seedlings are ready for transplanting. The seedlings are extracted from the nursery bed using a cutlass, and transplanting follows. Holes are dug at a triangular spacing of 2·5–3 m (8–10 ft) depending on the variety. The seedling is then buried in the hole and firmly pressed in position. When plastic bags are used, these are usually buried with the seedlings.

Weeding is very important, especially in the first 4 years of the plant's life. Immediately after planting, mulch is placed around the seedlings to conserve moisture, as cocoa cannot survive drought. The seedling is sometimes tied to a straight stick to encourage straight stem formation, and usually side branches are removed by pruning before the plant attains a height of 1·5 m (5 ft). As soon as vegetative cover is established all field shade plants have to be removed.

*Harvesting and Processing:* Maturity is 3–5 years after planting. Pods, carried on tree trunks or branches, start with a green or red-brown pod colour which ripens into yellow or orange. There are two harvesting periods in any one calendar year. The main crop is harvested between August and January, while the light or subsidiary crop is harvested between May and August. Harvesting is done gradually as enough pods become ripe. A very sharp-bladed knife is used to cut the pod off at its stalk attachment, taking care not to damage the cushion on the stem which will form the base of flowering and fruiting in subsequent years.

Harvested pods are broken up, again using a cutlass, and the seeds with their mucilage are poured out into a basket lined with leaves. There are 20–40 beans per pod. Average yields are 0·8–2 tonnes/ha (0·3–0·8 tons/acre) of dry beans.

*Fermentation* follows immediately after harvesting. The fermentation is done by heaping wet cocoa beans together, with regular turning to avoid excessive heat but to allow just enough to give the chocolate flavour. Each heap is usually covered up for 2–3 days after which it is turned and mixed, sometimes using wooden spades but more commonly by hand. Complete fermentation takes only a week. Nowadays, fermenting may be done in large wooden boxes with holes 1·3 cm (0·5 in) in diameter in the bottom to allow for draining of fluids. The tray method of fermentation is also used.

Fermented beans are then dried out on raffia palm mats on raised platforms by spreading out the fermented crops during the day, but only when there is sun. Drying lasts 7–14 days. After this the dried beans are picked free of impurities and sold to marketing board agents or directly to a government agent.

*Diseases and Pests:* In view of the importance of cocoa, extensive studies of diseases and pests have been undertaken, and control measures are being applied with vigorous routine on all cocoa farms. There are a number of virus, fungal and bacterial diseases of cocoa, the important ones of which will now be discussed.

*Swollen shoot* is a deadly virus disease of cocoa leading to loss of yield. It is readily transmitted by *meally bugs* helped by attendant *ants*. Symptoms include leaf mosaic, withering of the leaves. Sometimes die-back may result. Swellings on shoots and chupons of the plant indicate the advance form of the disease, hence the term 'Swollen shoot'. Control is by cutting and removing infected trees, or any showing early symptoms, followed by regular inspection and re-treatment. The latest method of control is to breed resistant or tolerant varieties.

*Black pod* is caused by the fungus *Phytophthora palmivora* Butl. The disease is evidenced by the presence of black patches on the cocoa pods and pod stalks. These patches are caused by the fungus feeding on the pod and, if uncontrolled, will cover the whole surface area of the pod. Infection is rampant and spreads easily during the rains, since the fungus is easily transmitted by water or wind. The beans also become black and worthless. Control is by the use of several fungicides (e.g. Perenox, Bordeau mixture) sprayed at regular intervals. Spraying is intensified during the rains and particularly when the pods are being formed. Removal and burning of diseased pods also help. Overhead shade as well as underbrush may also have to be thinned or completely removed to reduce dampness in the cocoa farm. Sanitation, drainage of excess water and good management in cocoa farms is imperative, and timely harvest minimises loss of yield. The use of resistant varieties is also to be encouraged.

Other minor fungal diseases such as *collar crack* (*Armillarea melles*), *brown root* and *white disease* are very prevalent in West Africa. These can be controlled by using fungicide sprays.

*Capsid* bugs attack cocoa, sucking the juice from the tip of the plant and so causing the death of seedlings and plant vegetation. Two important capsid bugs of West Africa are *Distantiella theobroma* and *Sahlbergella singularis*. Some *Helopetise* spp. also attack cocoa. The feeding habit of these capsids produces a lesion on the plant where the capsids have fed, especially on the young tender green shoots of the growing plant and later on the pods. These lesions are brown water-soaked masks about 1 cm (0·4 in) in length. Following the feeding the underlying tissues are destroyed, and the part of the plant above the point of attack withers away and dies for lack of food. Capsid attack is serious since it also gives rise to secondary infection by fungi. Heavy attack gives rise to browning of leaves and twigs as if there has been a fire rampage or 'blast'. Control is by the use of effective insecticides (e.g. Gammalin, Kokotae). Spraying is best begun when the capsids are known to start building up in order to prevent damage from the outset. This period varies, but at present much spraying is carried out during the dry season.

## 4.23 LATEX TREES

Rubber (*Hevea brasiliensis* Kunth)
Wild rubber is a feature of tropical equatorial rain forest. The cultivated crop derives from Para in the Amazon (Brazil), and for a number of decades now improved varieties (or *clones*) of the tree have been raised on plantations in West Africa.

The most important part of the crop is the *latex* which can be easily tapped from the stem at a height of 0·9–1·8 m (3–6 ft) above ground level. The latex is coagulated, pressed, sun-dried and smoked to produce rubber or crepe and used in local factories or exported.

Ecological requirements of rubber include an annual rainfall of not less than 200–300 cm (78–117 in) which is well distributed throughout the year and temperatures of 21–32 °C (70–90 °F). Rubber is also sensitive to altitude and will not yield well at heights above 600 m (1,970 ft). Best soil is one that is very fertile and free draining, with an adequate supply of water without being waterlogged. The response of the crop to magnesium, potassium and phosphorus is outstanding. The tree under ideal conditions can attain a height of 20–30 m (65–98 ft).

*Land Preparation and Planting:* The more common varieties planted in West Africa are the introduced *Malaysia* and *Indonesia* high-yielding clones. The land is cleared of undergrowth and holes 0·6 × 0·6 m (2 × 2 ft) in size are dug at spacings of 12–15 (40–50 ft) along rows which are about 30 m (100 ft) apart. These holes will carry the seedlings which must previously have been raised in nurseries.

In the nursery, the seeds are planted on sandy beds to induce rapid germination. Water is adequately provided, or, as an alternative, the nursery bed is made near a perpetually flowing stream and covered with mulch. The seedlings that emerge 8–10 days after planting are subsequently budded at 6–7 months with scion from high-yielding clones. If the budding is successful, the scion (bud) remains green and soon produces a shoot. At this stage the top portion of the original seedling is carefully and neatly cut off 8–10 cm (3–4 in) above the budding site so as to give the new shoot from the scion the best conditions for growth.

Field planting is done 3–6 months after budding depending on the growth of the new shoot. Transplanting is carried out preferably in the early mornings and only during the rainy season.

After transplanting, ring weeding around the trees is essential. Weeding along the rows helps to provide paths for tapping and collecting latex and for treating the trees.

P–K–Mg fertilisers are very conducive to good growth and development. Dead, poorly developed or infected trees should be replaced as soon as they are detected.

*Harvesting and Processing:* After 5–7 years from planting the tree is ready for *tapping*, by which the latex is extracted from the tree. Tapping can continue until the tree is 40 years old, when it should be replaced. Tapping is a skilled job. The bark of the tree is removed, and a cut is made from left to right down half of the tree at 30° to the horizontal. A collecting spout is placed at the end of the cut, which conveys outpouring latex into a cup at the end of the spout. The channel is cleared daily, with new incisions made moving downwards semi-spirally daily.

It is important that the cambium layer of the tree just below the bark is not in any way damaged; otherwise the tree may be permanently injured, and regrowth of tissue may prove impossible. If the cambium is intact, the tree will regenerate. The tapped surface will be adequately replaced, and the tree will be ready for another set of tapping the following year.

Latex is a white exudate from the tree and must be collected before sunrise to avoid coagulation. Usually some anticoagulants are added to the latex as it is being collected in the cups on the trees.

The yield of latex varies depending on: (a) variety planted; (b) climatic conditions (more latex is produced in the wet season); (c) the tapping method and efficiency of tapping; (d) the soil conditions; and (e) style of cultivation (plantation cultivation yields more than peasant cultivation as mixed cropping). Yields of 1·5–3 tonnes/ha (0·6–1·2 tons/acre) are possible on plantations, while the average yield on peasant farms is about 1 tonne/ha (0·4 tons/acre).

Processing of latex is done immediately after collection. The latex is poured into a big container (*cistern*), filtered and then mixed with weak acid (usually formic acid) to *coagulate* the clean latex. The coagulated *rubber* is washed through, cut into sheets, pressed, rolled and sun-dried before being smoked for 4–5 days in specially prepared ovens. The smoke house is first kept at 49 °C (120 °F) to allow for a slow drying to start with, then later raised to 60 °C (140 °F).

Another processing method, especially with rejected or field-coagulated latex is by *crepeing*. The latex is soaked in water for a number of days and then pressed through a heavy metal drum. These crepe sheets are pressed thin, sun-dried and smoked in the same manner as rubber sheets.

After smoking, rubber sheets and crepes are graded and packed into bundles before being transported to factory or store. Export of rubber sheets from West Africa produces appreciable income for both growers and traders.

*Diseases and Pests:* Three important fungal root diseases associated with rubber are *white rot* (*Fomes lignosus*), *brown rot* (*F. noxius*) and *collar rot* (*Ustulina zonata*). These fungi are capable of attacking other trees, and deadwoods serve as secondary hosts. The affected crop loses yield and may die.

The tapping panel can also be infected by several fungi. Even in the smoke house, if correct temperature is not maintained, fungi can damage wet rubber sheets or crepe, thus reducing their market value.

Control is by the use of fungicide. The root diseases can be treated with partial success. To treat, the roots are first unearthed (exposed) and scraped carefully free of the fungal tissues, and then fungicide is applied. Once the fungi are killed, the roots are again covered. Careful regular inspection helps to reduce losses. The use of resistant varieties as root stock is recommended.

## 4.24 EDIBLE FRUITS AND NUTS

There is an impressive number of tropical fruits and nuts grown all over West Africa for local consumption as well as export. The more important crops include mango, guava, citrus, banana, pawpaw, pineapple, avocado pear, cashew, date, coconut, almond etc. to name a few. But perhaps the most common are the citrus fruits, avocado pears, bananas and pineapples which are all cultivated as plantation crops all over West Africa. Part of the harvests from these plantations are fed to local processing factories for juices, marmalades and other extracts in cans and bottles.

Citrus (*Citrus* spp.)
The citrus fruit trees of West Africa are evergreen shrub-size plants and include sweet orange (*C. sinensis*), grapefruit (*C. paradisi*), tangerine (*C. reticulata*), lemon (*C. limon*), lime, sour orange and a few others. The ripe fruits can be eaten raw. The rind is useful in the preparation of etheral oils, cattle feed and marmalade. A few countries in West Africa export grapefruit to Europe.

Citrus plants will tolerate a very wide range of temperature and climate. Different varieties, however, are more tolerant than others. But at temperatures below 15·5 °C (60 °F) the tree will not survive. Soil requirement is more important. It has to have a deep topsoil which is rich in minerals and humus, free draining and loamy. A regular supply of water is essential. Hence an annual rainfall of about 100–200 cm (39–78 in) is required and spread evenly throughout the year.

*Land Preparation and Planting:* The requirements and land preparation for citrus are similar to those of the other economic trees discussed earlier. Holes are dug at spacings of 4·5–7 m (15–23 ft) depending on variety and size of plant. Fertilisers (nitrogen, copper, zinc, iron and trace elements) are added for optimum yield.

Usually, citrus plants are propagated by grafting or budding. The seeds are first raised in trays which are kept under shade and later in nursery beds where the budding or grafting is done. They are finally taken to the field when the budded seedlings are mature enough. The seeds germinate to provide later the root stock of the citrus plant. In view of this, only the varieties resistant to a number of root diseases are used for the root stock, while the higher-yielding varieties are used for the budding stock (scion).

Seeds are planted in trays or seed boxes at 4–5 cm (1·5–2 in) intervals, and 6–8 months later the seedlings are transferred to the nursery where they are planted 1·2 m (4 ft) apart. In the nursery bed, staking is done to train the seedlings to have a straight trunk. All side shoots are pruned until the plant has reached a height of 1–2 m (3–6·5 ft).

Final transplanting to the field is done in the early morning, between May and July during the rains. The long tap root of the seedlings is very sensitive at the tip where the root cap is. Care must therefore be taken to avoid any damage to the root. All leaves are removed by cutting off at the stalk of each leaf.

After transplanting and mulching, the field can be sown to cover crops or grass to check erosion. Grazing of livestock is not recommended on citrus farms, principally because of the thorns on the citrus and partly because livestock may damage or eat the young plants. Grasses on citrus farms should, however, be slashed regularly, and the cut herbage can be stall-fed to livestock. Clean weeding around individual plants is essential.

*Harvesting and Storage:* After 4–5 years in the field the citrus trees start to bear fruits in commercial quantities. To harvest, the tree is climbed, and the fruits are plucked and put (*not thrown* to avoid bruises) into a jute bag or a basket lined with jute to avoid damaging the fruits.

Harvesting usually starts by October, and fruits are harvested as they ripen, i.e. when the colour changes from bush green to orange, yellow or yellowish red depending on the variety.

Yield varies according to variety planted, soil, climatic conditions and farm management. Average annual yield per tree is 0·6–1·5 tonnes/ha (0·2–0·6 tons/acre) for sweet orange and relatively less for grapefruit. Yields of about 18–25 tonnes/ha are possible with improved varieties.

Harvested fruits are either used or sold immediately or delivered to the factories. Occasionally it may be necessary to store harvested fruits until there are enough for sale. Two very important points to be borne in mind with citrus are: (a) avoid heaping fruits on one another; and (b) always store on the floor in cool rooms or in open drawers where the fruits are spread out.

*Diseases and Pests:* Two important fungal diseases of citrus are *root rot* and *brown rot* which are caused by *Phytopthora* spp. When attacked, the roots tend to rot, and gummy materials are exuded from lower portion of the citrus stem. In West Africa the name *'gummosis'* (production of gum) is generally applied to these diseases. Control is by the use of fungicide. The roots have to be exposed (unearthed) and scraped free of the fungi before fungicide is applied. The use of root stock from plants resistant to these diseases is strongly advised. Keeping citrus farms healthy and free of undergrowth also helps.

Virus diseases such as *triteza* and others are also common on citrus plants in West Africa. When attacked, the plant may suffer a die-back, in which case the whole plant should be uprooted and burnt. The aphids that feed on citrus plants constitute a common vector of this disease; hence control by spraying citrus plants with insecticide is advantageous.

Pests such as *bugs*, *mites*, *aphids*, *grasshoppers*, and larvae of *butterflies* and *moths* all suck, eat and damage leaves, stems and fruits. Control is by timely harvesting and spraying with insecticide.

## 4.3 GRASSES AND COVER CROPS

### *Pasture Establishment and Maintenance*

Pastures comprising either grasses alone or grass–legume mixtures are established for a number of reasons: (a) for livestock feed as grazing;

(b) to check soil erosion; (c) to adorn the landscape (e.g. lawn near a house); and (d) for sport (e.g. golf, lawn tennis). The use of pastures for livestock feed is the most relevant here.

Grasses are essentially herbaceous plants with hollow stems except at the joints or *nodes*, hence referred to as *culms*. Flowers of grasses are usually hermaphrodites with the male and female inflorescences arranged as *spikelets*. However, the propagation of grasses is through the *runners*, which are partially underground stems, and also by seeds.

In West Africa the establishment of pasture is fairly difficult, although in the savanna belt grasses grow abundantly in their natural state. The land is prepared by clearing or burning and cultivated with hoes or ploughs, and the seeds are broadcast or, more usually, runners are buried in the earth in rows not more than 50 cm (20 in) apart depending on the grass variety. Nitrogen fertiliser and mixtures of NPIC can be applied to stimulate rapid growth. Once the pasture is established, weeding is no longer necessary, even though it is vital to be on the lookout for weeds, some of which can be very dangerous and poisonous to livestock.

Routine maintenance comprises the following:

(1) Do not overgraze the pasture; practise rotational grazing.
(2) Remove weeds from time to time.
(3) Apply fertiliser and water as the need arises, especially in the dry season.
(4) Do not allow the grass to grow too hardy; therefore cut back regularly to stimulate the growing of young succulent shoots and leaves.

*Grass Pastures*

In West Africa, grass is part of the natural vegetation of the savanna belt. However, the most common of the grasses used in the establishment of pastures are discussed briefly below.

(1) *Guinea grass* (*Panicum maximum*): This is a perennial grass widely distributed in West Africa. It is easily propagated by seed or by stem cuttings called *stolons*. It is tufted in appearance and grows luxuriantly to a height of about 3–5 m (10–16·5 ft). It can be easily cut and stall-fed. It can also be cut and baled as hay, and poor quality hay can be used as livestock bedding for farmyard manure (FYM).
(2) *Elephant grass* (*Pennisetum purpureum*): This grass is perennial too and grows to a height of 4·5–5 m (15–16·5 ft) like guinea grass, to which it shows similar ecological affinity. Both are propagated in a similar manner, i.e. vegetatively. Elephant grass is

relatively more resistant to drought than guinea grass. Best yields are obtained when the grass is cut young. The more hardy it becomes, the poorer is the feeding quality.

(3) *Giant star grass* (*Cynodon* spp.): Several species of this grass are cultivated in West Africa. *Cynodon dactylon* is a virile grass which by virtue of its spreading habit becomes quickly established and likely to smother any weed or plants grown with it even as legume mixtures. It is difficult to eradicate but useful in keeping down weeds and in checking erosion on river banks and bad slopes.

(4) *Gamba grass* (*Andropogon gayanus*): Gamba grass is indigenous to the savanna belt where it grows extensively in its natural form. The natural grassland having *Andropogon* spp. can be improved through careful tillage, removal of weeds and the application of fertiliser and water. It can also be easily established using seeds or stolons. It grows to a height of 2–2·5 m, (6·5–8 ft), depending on variety and maintenance.

(5) *Rhodes grass* (*Chloris gayana*): This is another fairly tall, perennial, drought-resistant grass that is easily established through broadcasted seeds. It is cherished by all classes of livestock and very responsive to fertiliser.

### *Fodder Cane and Silage*

Some grasses are cultivated as fodder in that they are cut and stall-fed to stock. They are not normally grazed. Young fodder is cut at its optimal growth, i.e. before the stems harden, then made into hay or silage and fed to stock in the dry season when grass herbage is usually poor or non-existent.

*Gamba grass* (*Andropogon gayanus*) is one of the most popular fodder grasses. When being cut into hay, the optimum conditions of cloudy weather, high temperature, high humidity and heavy rainfall help.

### *Grass–Legume Mixtures and Cover Crops*

Grasses and legumes when established together in varying proportions are regarded as good pastures with a balanced nutritional constitution. Both are usually established through seeds which are broadcast or drilled. Fertiliser may also be added at the early stages to encourage rapid establishment. Nitrogen fixation by the legume crop in the mixture helps the pasture to maintain good vegetative growth all year round.

The more common grass–legume mixtures in West Africa are: (a) centrosema with giant star grass; (b) *Stylosanthes guyanensis* or *S. gracilis* with guinea grass or rhodes grass; and (c) *Stylosanthes* sp. with *Pueraria* sp.

All legumes are propagated by seeds and form pods if allowed to mature. But in pastures, growing to seed is not advisable since at this stage the legumes are hardy, unpalatable to stock and difficult to digest. Apart from being a component of good pastures, legumes are also used as cover crops, for shade and, more especially, to check erosion.

These grass–legume mixtures can be cut and stall-fed as green fodder or made into hay or silage depending on the climate and ratio of pastureland to stock. Grass–legume mixtures make very good hay and silage, but an expert is needed to achieve the best combination.

*Part Three*

# THE PRODUCTION AND MANAGEMENT OF FARM ANIMALS

*Chapter 5*

# Science and General Principles of Livestock Production and Management

## 5.1 GENERAL PRINCIPLES OF LIVESTOCK MANAGEMENT

### 5.11 MANAGEMENT SYSTEMS

In a sense, we can speak of three types of stock on any farm. There is the *dead stock*, a general term used for buildings (e.g. houses and pens), constructions (e.g. dams), machines (e.g. tractors, grinding mills), equipment (e.g. spraying pumps) and tools (e.g. hoes, cutlasses). The second category of stock is *crop stock*, which includes seeds, crops planted and crops in store. The third and final category is *livestock*, which includes cattle, sheep, goats, pigs, dogs, poultry and fish. Fish have generally not been regarded in the conventional sense as farm animals, but they are now; this is because of the growing importance of fish and fish products for the supply of protein to the inhabitants of the countries of West Africa. In addition, fish ponds and inland water fishing in rural areas are becoming more and more important in the efforts of many governments to improve both the cash income and living standards of many of their people, particularly protein in their diets. And fish is a very good source of protein.

The management principles for livestock, to be fully understood, must commence with and be related to the type and use of farm animals, fish and their products. The major factors to be considered in management are: (a) type and use of farm animals and fish, (b) selection, (c) feeding, (d) housing, (e) hygiene and sanitation and grooming, and (f) reproduction.

Four main management systems for farm animals and fish that can be identified in West Africa are the nomadic, free range, intensive and semi-intensive systems.

*Nomadic System*

The first and oldest system of keeping livestock in West Africa is the nomadic system. In this system, sometimes referred to as *Transhumance*, the man and his livestock move from place to place in search of water and pasture for the animals. With fish, the fisherman similarly moves from one water mass (rivers, springs and lakes) to another, trapping the fish with his hands, baskets and spears or by poisoning or using nets. This management system is very rudimentary, simple and cheap, but it is crude and very unprofitable because it does not include giving something back for the future on the part of the livestock man or fisherman. Very soon, all the stock may perish, as in the case of drought, or the fish population may be completely exhausted. In effect his very survival is threatened, not to speak of the starvation, malnutrition and diseases that follow from time to time at an epidemic rate.

*Free Range System*

The next stage to the nomadic system is the free range system. In this system the livestock man (or farmer) is settled in one area or village. He builds himself and family a house, but his livestock still roam about, either by themselves (e.g. poultry, dogs, pigs) or with a herdsman (e.g. cattle, goats, sheep). Occasionally these livestock get to feed on wastes from the farm or kitchen. In general there is very little care, if any, and additional food is rarely provided. In this system the cost in terms of labour and cash (or investment) is very very small. As a consequence, production from the stock is small and usually of poor quality; losses are high, and the reproduction rate is poor because of the many hazards the livestock have to face.

Poor, unthrifty and unscientific as this system is, the majority of livestock (nearly 90 per cent) in West Africa are still kept in this way. Consequently the development of livestock in West Africa has been slow, with the result that the protein supply in the diets of most West Africans is poor.

*Intensive System*

This system can be regarded as the extreme opposite of the nomadic system. In this case the livestock are kept under the most intensive and highly scientific management. They are housed indoors all the time with the temperature of their environment controlled at optimum level: feeding is scientifically balanced and precisely fed, for the best use of and maximum output from the livestock. Movement of the animals is restricted to the minimum necessary, and every stage of the animals' lives is controlled and stage-managed for the maximum benefit of the animals and their keepers.

This system is very expensive and labour-intensive. Therefore it can only be applied to livestock of good performance breed which have been highly bred to respond to these types of intensive treatments. Because of the high costs involved, this practice (e.g. battery system for poultry, stall feeding for cattle, sheep and goats) is not within the reach of many poor peasant farmers in West Africa. Hence in West Africa this system is confined largely to universities, research institutions, government farms and a few highly commercialised farms where funds are available and scientific knowledge is developed and practised. With the growing awareness of the merits of improved agriculture, this system is being gradually nurtured, introduced and encouraged for some private farmers to undertake. The battery system for poultry, for instance, is now common with most poultry keepers near most city centres of West Africa.

*Semi-intensive System*

This system can be regarded as the middle stage between the free range and the intensive systems. The animals are housed part of the time, food is supplied regularly, and sometimes the animals are allowed out to graze and fend for themselves in specially allocated pastures or pens. In a sense, it strikes a balance in very many aspects between the too expensive and highly technical intensive system and the relatively less productive but cheap free range system. In terms of cost, labour requirement and productivity, this system seems to be optimal at the current stage of scientific development, technical know-how and available funds for livestock development in West Africa.

## 5.12 USE OF FARM ANIMALS

*As Security*

To most people who keep livestock in West Africa, animals provide a source of security and comfort. It is more than an investment or wealth; it is a heritage. To understand some of the attitudes of these livestock men and to appreciate their usual reluctance to sell or dispose of even very sick and old animals, it is important to take this into consideration. Therefore, any effort to improve livestock husbandry and management in West Africa must, of priority, give consideration to this important wealth consideration which induces people to keep livestock as a source of security.

*For Cash*

Even when peasant livestock keepers refuse to sell the sickly and haggard-looking ones of their stock, for reasons mentioned above, they have little or no hesitation about selling young stock or livestock products for cash. Milk, eggs and meat are all readily sold. The cash from sales forms a major source of income for them. Fishermen too depend nearly exclusively on the money from the sale of their catch for buying their other needs, such as clothing, school fees and other family expenses.

With commercial agriculture, cash income is the sole aim of such undertaking. Livestock and livestock products are disposed of solely for their cash returns. Similarly, large commercial fishing organisations, like any other business, are out solely for profit and sell their products and by-products (fish and fish meal) for as much as possible.

On the whole, farm animals are kept principally because of their ability to convert grasses and pasture herbage as well as farm wastes into cash. Similarly, fish convert marine small life and other wastes into fish and fish products which are later sold for cash.

*Food*

Livestock products such as meat, fish, milk and eggs are the best sources of protein. Farm animals and fish therefore provide the farmer, his family and others with a good supply of protein in their diets.

Livestock can feed on grasses and household wastes and convert these into useful food products. Milk, a unique product of the mammalian family of the animal kingdom, is a most important animal product from

Table 5.1 *Composition of milk from different animals, in percentage by weight (i.e. grams per 100 g, or ounces per 100 oz).*

| *Mammalian Species* | *Water* | *Fat* | *Lactose* | *Protein* | *Minerals* |
|---|---|---|---|---|---|
| Man | 88·5 | 3·5 | 6·5 | 1·3 | 0·2 |
| Cattle (cow) | 87·5 | 3·3 | 4·6 | 3·5 | 0·9 |
| Sheep | 83·0 | 6·5 | 4·3 | 5·8 | 0·4 |
| Goats | 87·0 | 4·5 | 4·3 | 3·8 | 0·4 |
| Pigs (sow) | 83·0 | 6·5 | 4·0 | 6·0 | 0·5 |

cattle, sheep, goats and man. It has a high proportion of protein, particularly the essential amino acids (e.g. lysine, tryptophane), as well as important minerals (e.g. calcium, potassium, sodium chloride). Milk composition for different members (species) of the mammalian group is given in Table 5.1. In addition to fresh milk, other important milk by-products (e.g. butter, cheese, ghee, skim milk, milk powder) are extremely good food.

Some by-products of livestock products (e.g. fish meal, bone meal, blood meal) are useful ingredients of livestock rations, supplying protein and minerals in significant amounts to animals fed on such rations.

*Work and Traction*

Work-bulls, donkeys and horses in most countries of West Africa are used for ploughing, carrying loads and pulling (e.g. in traditional sugar-cane crushing). The use of animals for work and traction is, however, not a very common practice in the tropical equatorial zone (south) of West Africa, principally because of the prevalence of the tsetse fly, which carries the organism *Trypanosoma*. This organism causes the deadly disease called trypanosomiasis (sleeping sickness) in both cattle and man. This disease has been the major limiting factor in the extensive rearing and use of livestock, particularly cattle, in the coastal areas of West Africa.

*Manure*

One of the advantages to be derived in mixed farming is the supply of a relatively cheap form of manure. From livestock the farmer derives farmyard manure (mainly from cattle) and compost manure (from pigs, poultry, fish wastes, sheep and goats). Manure is useful for the physical improvement of soil and for supplying nutrients, particularly humus, to the soil. Manure is supplied in the form of prepared farmyard manure (FYM), or as compost, or through kraaling.

*Farmyard Manure* (FYM) is a product of three activities by the animals. Straw bedding is put in the animals' pen, and the animal supplies dung, urine and physical pressure on these. The urine soaks the straw, and the sheer weight of the animal trampling and lying on the dung and urine causes partial breakdown to take place in all the waste products, producing as the end result a first-class balanced manure containing nitrogen, phosphates and potassium compounds (NPK). The manure is subsequently removed from the pen and kept in a pile in a cool place away from direct heat and sunlight to conserve the rich and highly volatile and gaseous nitrogen compounds in the manure. Largely because of the straw content in FYM, it helps in the structural improvement of the soil in addition to supplying nutrients and keeping a high level of micro-organisms in the soil. Micro-organisms in the soil help to continue decay activities, to release more nutrients to the soil. Aeration of the soil is also improved.

*Compost manure* is made by the farmer. Other farm animals (e.g. poultry, pigs, sheep, goats, fish (fish manure)) produce a sizeable quantity of dung and wastes which can be collected and composted along with bedding and food wastes from the pen to produce a high quality manure. Poultry faeces, for instance, contain a high proportion of calcium (lime) and phosphorus which is particularly good for acid soils.

Compost can also be made from farm and kitchen wastes and crop residue, with or without adding the livestock residue. Compost making is a very simple process. The most popular method of compost making in West Africa originated from India and is usually referred to as the Indore system of composting. In making compost, one can either start from the flat or dig a pit about 1·8×0·9 m (6×3 ft) in area and 0·9 m (3 ft) deep. The pit could be slightly bigger or smaller depending on the available waste or compost material. In dry areas compost materials are better put into a pit, but not so in wet areas because of waterlogging.

First the materials are laid in heaps and compressed to not more than 15–20cm (6–8 in) thick by pounding to reduce air spaces. The dry materials layer is made wet with water or urine, and this layer is followed by a layer of ash/urine/dung as a 'starter', made up from already decomposed material. The starter introduces the bacteria and other micro-organisms required for starting decomposition in the dry materials below. The starter is spread evenly over the first layer; water is then sprinkled over, and a next layer – again 15–20 cm (6–8 in) – of dry compostable materials is added, followed by a repetition of the dung/urine/composted or ash innoculant or starter and water. The layers follow one another until the pit is filled and a dome-like rise is established about 15–20 cm (6–8 in) above ground level. The pit is left for 3–4 days for decomposition to take place.

While the first pit is undergoing decomposition, an adequate amount of water is supplied twice daily. A similar pit is dug next to the first one in readiness to receive the turnover from the first pit. Approximately 2 weeks after the filling of the first pit, the whole material is turned over into the second pit; it is compressed, and water is added. The rate of turning and the amount of water added will depend on the speed of action that has taken place in the previous pit. After turning over the contents of pit no. 1 into pit no. 2, the pit no. 1 is now refilled as previously with farm wastes, dung, etc. in the same order.

Two weeks after the first turning, a third pit of similar dimensions is dug, and the contents of pit no. 2 are transferred into pit no. 3. Similarly, the new contents of pit no. 1 are again transferred into pit no. 2, and pit no. 1, now empty, is made up again as previously. Thus the whole cycle continues. The turning from pit to pit in regular order continues until all available materials are composted or until the decomposition is complete and the compost is ready.

The compost may not be ready for use immediately and may have to be stored. A shed has to be provided, and the stored compost has to be protected from rain, wind, heat and direct sunlight. It should be covered, preferably with palm fronds or light jute bags.

The quality of the compost is influenced by three things: (a) the quality of the materials being composted – the richer the better; (b) the rate of decomposition; and (c) the storage – poor storage will lead to loss of nutrients, especially the nitrogen compounds, which are easily lost either by leaching if the compost is exposed to rain or by oxidation if exposed to heat and/or sunlight.

Livestock also provide manure to farms through *kraaling*. Usually after field crops (e.g. maize, guinea corn, millet, groundnuts, cowpeas) have been harvested, the haulms and crop residues provide good hay for the animals. Livestock can then be kraaled on these farms for two or more days, during which time they feed on crop residue while at the same time depositing their dung and urine on the farms. These animal wastes are subsequently ploughed in as good quality manure. This kraaling can also be practised on pasture land just before reploughing.

*Subsidiary Uses*
In addition to the five major uses of livestock discussed above, livestock are also useful, perhaps to a lesser degree, for the following: (a) hides and skins (e.g. cattle, sheep, goats); (b) hunting, sports and games (e.g. dogs, horses, polo ponies, donkeys); (c) pets (e.g. monkeys, cats, dogs, some less harmful reptiles); (d) zoos, for educational purposes; and (e) by-products – for instance, horns and feathers are used for decorations, bones for minerals in feed mixtures for livestock and hoofs for gums, glues and several other uses.

Generally, the good management of farm animals helps to make sure that farm animals and fish are given the most beneficial conditions and environment so that they can produce their best. In this way these animals are put to their best use. Most of the uses discussed above are for the benefit of their owners in the first place. Therefore farmers have much to gain by trying to achieve on their farms a very high standard of management for their livestock.

## 5.13 SELECTION OF GOOD ANIMALS FOR THE FARM

The best start of good livestock management lies in the selection of good animals for the farm. Three basic principles or factors influencing the selection of any farm animal are: (a) the type of product or use the animal is for – i.e. whether for breeding or only for producing milk, meat, eggs or fish for sale; (b) the level of management to be adopted – whether free range, intensive or semi-intensive; and (c) the cost of the animal, especially with respect to what the livestock keeper or fisherman can afford.

In every case, however, the following seven major factors must be taken into consideration when selecting a farm animal, irrespective of the use, management and cost.

*Heredity*
The parents of the farm animals, especially the sire (male), has to be well known and must be of good stock. Well-bred animals are said to be of *proven stock* when the performance or productivity of their

parents is very economic and their products are of high standard. Always select from a proven stock. This is very important if the animal is to be used for breeding.

*Health*

The animal to be selected must be healthy. To a certain extent, the health condition of the animal can be judged by examination of its specific parts:

(1) *Eyes:* Should be clean, clear, free from mucous, with no lacrymation (tears) and not bloodshot.
(2) *Nose:* Should be fairly moist without being too wet, free from any discharge or wound, not swollen or infected, free from ticks and other parasites and pests.
(3) *Mouth, tongue, teeth and gums:* Should be blood filled, moist without discharge; teeth should be strong, well formed and unbroken; internal mouth parts should be free from swellings, with no lacerations, wounds or cuts, or any infection.
(4) *Anus:* Should be free from swellings or cuts; the dung and urine should be of the right texture and consistency, i.e. neither too hard nor too watery and not tainted with mucous or blood.
(5) *Legs,* particularly the hoofs: Should be well formed and free from infection, swellings, cuts and bruises.
(6) *Skin:* Coat should be shiny (glossy), of right colour and covering, and free from cuts, bruises or other blemish, particularly ticks and other parasites and pests.
(7) *Reflexes:* Should be good. (Reflexes are defined as automatic natural reactions resulting from nerve response to external stimuli, such as closing of eyes at the approach of something to the eyes, jerking of the knee in response to being hit, and so forth.

Resistance to deadly local diseases is also very essential.

*Age of the Animal*

The younger the animal the better, and probably the cheaper. It is helpful to have an animal at a tender age. This gives the breeder the opportunity not only to train the animal carefully but also to give it the best of opportunity aimed at maximum production later.

*Fertility and Fecundity*

The ability to reproduce is partly a breed factor, but it is also affected by heredity and the environment under which the animal was raised. Where an animal is to be used for breeding, it is important to select one from a well-known herd of very fertile and fecund animals.

*Conformation or Body and Build*
Again the build or conformation of any animal will be along the lines of the breed. Conformation is therefore a hereditary factor, but it is largely influenced by the environment (food, health and care) under which the particular animal was raised. It is advisable to select an animal on the basis of good conformation. This can be judged by examining the essential parts of the animal in relation to the use to which the animal will be put.

(1) *Dairy animal (cow):* Should be of slender body and low body weight, but the udders (milk factory) have to be big, healthy and free from infection, with pronounced and open teats.
(2) *Beef animal:* Should be full of flesh (thick muscles) around the rump, forelimbs and chest; the legs are preferably short, the rump square and the back long and broad.
(3) *Work animal:* Needs long legs, a thick neck and broad shoulders for carrying the plough; it must be very docile, of good temperament and very strong.
(4) *Egg-laying bird:* Should be of a good breed with a reputation for a high egg-bearing annual average, of good body weight relative to age and resistant to local diseases.

The main features of various types of livestock are illustrated in Figures 5.1 to 5.5.

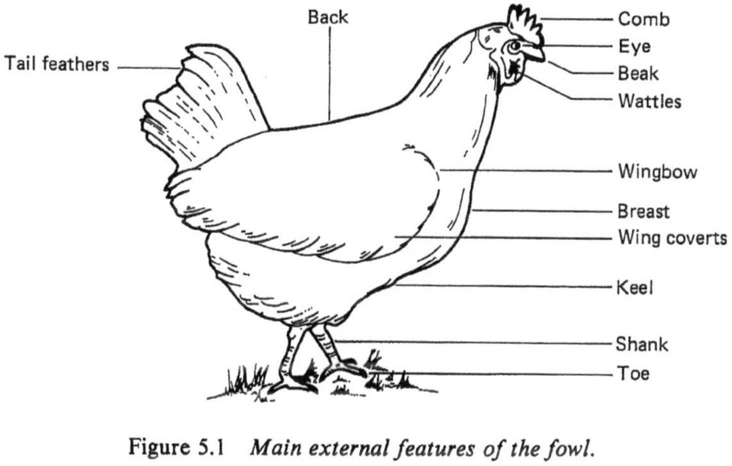


Figure 5.1 *Main external features of the fowl.*

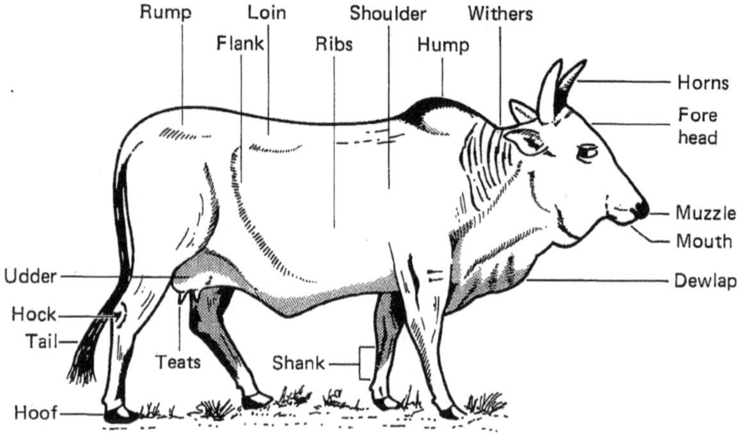


Figure 5.2 *Main external features of a cow.*

Figure 5.3 *Two common fish types in West Africa.*

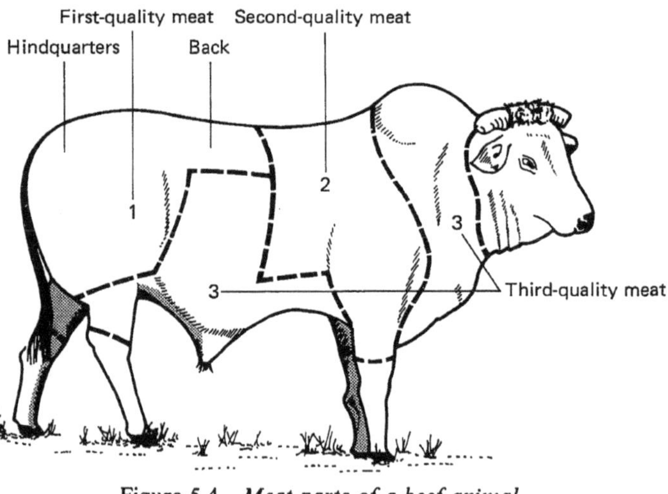


Figure 5.4 *Meat parts of a beef animal.*

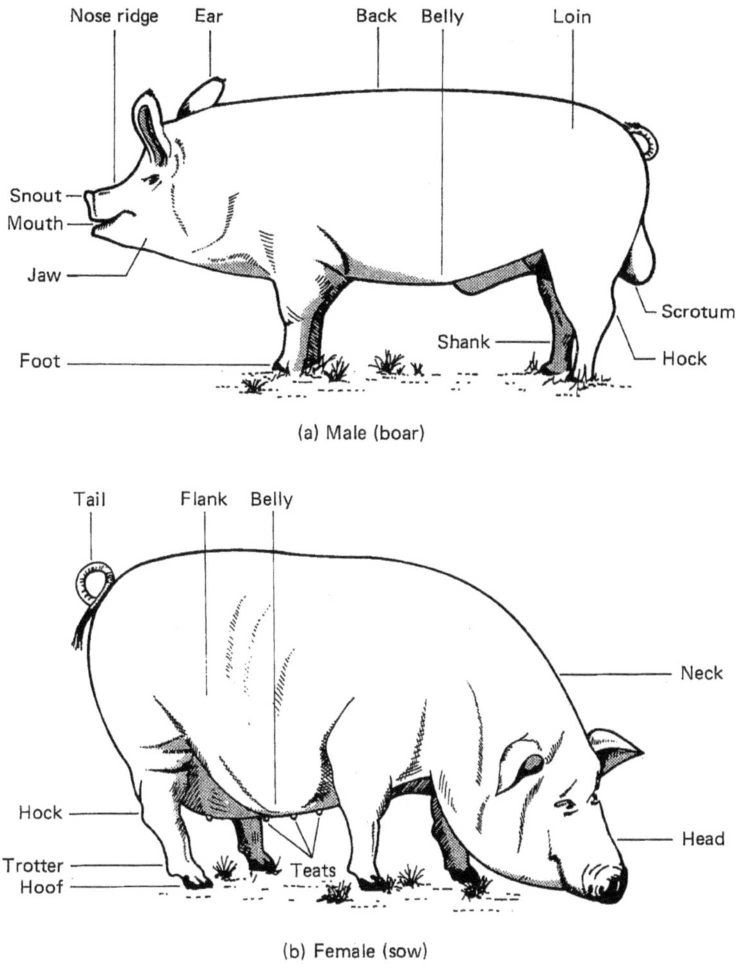


Figure 5.5 *Main external features of a pig.*

## 5.14 FEEDING

Animals, like human beings, have to eat. Well-fed animals grow faster and bigger and produce more and better products (e.g. milk, eggs, meat); they can be sold at a higher price too. Feeding the animals well helps to keep the animals healthy and more resistant to diseases and pest attack. They reproduce better, and the offspring are strong with a good chance of survival. On the whole, animals eat for a number of reasons: (a) for the day-to-day *energy* needed for breathing (respiration), standing, sitting, walking – hence maintenance ration; (b) for *body building*, i.e. growing meat, bones, blood and hair/feathers; (c) for *work*, i.e. drawing ploughs, carts and other loads; (d) for *producing* meat, milk, eggs – hence production ration; and (e) for *combating diseases* – minerals and vitamins are needed for building hormones, anti-toxins, as well as extra white and red corpuscles of the blood.

Different animals have different kinds of organs for digesting the food they eat (*digestive system*). They can therefore be classified according to their *alimentary canal*.

*Fish* as a group have a simple alimentary canal. All foods (e.g. bacteria, seaweeds, smaller sea animals, wastes, other pieces picked up) are ingested with water. These pass through the mouth into the simple stomach where digestion takes place.

*Birds* have no teeth but can pick their food with their beaks. Thus they are able to feed on grains and semi-solids (mash) and to tear at soft plants and succulent tissues (greens).

But in the case of mammals, particularly the *quadrupeds* – i.e. the four-footed mammals – there are three or four important groups.

(1) *Carnivores* (e.g. cats, dogs) feed mainly on other animal tissues because their digestive systems cannot handle plant tissues.
(2) *Omnivores* (e.g. pigs, man) eat both plant and animal tissues.
(3) *Non-ruminant herbivores* (e.g. horses, donkeys, rabbits) feed mainly on succulents, plant roots and seeds and little on herbage.
(4) *Ruminant herbivores* (e.g. cattle, sheep, goats) feed mainly on plant herbage (grasses and legumes) because of the special nature of their digestive system. In particular, the stomach of the ruminants is divided into several segments or apartments, each specifically adapted for handling leaves of grasses, shrubs and other herbage. They can also handle other foodstuffs, of course. These segments – the *rumen, reticulum, omasum* and *abomasum* (see Figure 5.7) – can retain the herbage during the digestive process, to allow different enzymes to act on the food, aided by bacterial and other micro-organisms, before the partly digested food is pushed from one section to another.

Feeding of the *young animal* is a very delicate management affair. There is a more detailed discussion on this later in section 5.23, but it is essential to mention now the few major points to consider in feeding young animals.

(1) Young animals should be fed in small quantities three or four times at regular hours daily.

(2) The food should preferably be in liquid form and high in protein that is easily digested (milk is ideal if not too expensive).
(3) In the case of mammals, the young ones should be fed either directly or indirectly from the mother's milk at least in the first 2 weeks, to allow for the intake of the *colostrum*, i.e. the milk containing antibodies from the mother.
(4) Always feed at body temperature as far as possible.
(5) Never feed stale foods.
(6) When the young are put on solid foods, it has to be balanced and preferably fortified with vitamins and minerals, and given in small quantities to start with.

## 5.14 HOUSING AND GROOMING

### *Housing*

Housing farm animals is an expensive item of management and is second in importance only to feeding. Animals are housed to protect them against wild beasts and thieves. It is useful also to keep animals protected against too strong wind, rain, too strong direct heat of the sun, disease-causing organisms and pests.

Several factors have to be considered in housing farm animals:

(1) *Type of animal:* Birds not only require smaller space but will also do reasonably well in temporary structures which can be easily kept clean. But with cattle, horses and pigs the buildings have to be strong, solid and well ventilated.
(2) *Type of production:* The type of production for which the animals are being kept is important. Where the product is expensive and cleanliness is important, as in the production of milk, butter, cheese and eggs, the buildings must be such that they can be easily and regularly cleaned. All necessary precautions must be taken to avoid contamination of the product and to maintain the health of the animal.

    Animal houses do not have to be very expensive. Buildings made of local materials (e.g. wood, earth, straw) can often be quite adequate. The golden rule is that if the expenditure does not lead to increased revenue which will cover the extra expenditure, then it is not profitable to spend such money.
(3) *Type of management:* This is an important item, since for some types of management (e.g. the free range system) housing is not necessary. Similarly, with the semi-free range system elaborate housing is not important. However, when animals are kept under the intensive system the housing situation is a key to successful business and profit.

On the whole, the essential features or conditions for good housing include the following:

(1) The animals are protected from rain, heat, wind, pests and parasites.
(2) They lie on a dry place (e.g. dry clean straw).
(3) Avoid draught.
(4) Keep off flies as much as possible.
(5) Avoid overcrowding; a cow, for instance, needs about 5–6 $m^2$ (54–64 $ft^2$) of floor room.
(6) Remove dung and manure from the house as often as possible.
(7) Avoid sharp edges and holes in buildings as these may cause accidents and hurt the animals.
(8) Always surround animal houses with a strong fence to provide a space for exercising the animals. Wherever possible, the fenced area (*paddock*) should have a few trees to provide shade.
(9) As far as possible, feed troughs and water troughs should be provided within the building as well as in the paddock.
(10) Always provide a separate building (or area) where sick animals can be isolated as a control measure against the spread of disease.

*Grooming*

Grooming covers activities such as brushing, washing and cleaning of all parts of the animal body. Grooming farm animals is important for many reasons:

(1) Grooming keeps the animals free from ectoparasites, which are easily brushed off if grooming is done properly.
(2) Grooming encourages the taming of the animals in that they get used to being handled by people (caretakers).
(3) During grooming, wounds, cuts and injuries are detected from the reaction of the animal, especially when the affected parts are touched; sick animals and animals on heat are also quickly detected.
(4) Grooming can also be used as a means of keeping the animal warm, particularly if the animal has been out in the rain or cold weather.
(5) Grooming allows for other operations such as dehorning, tattooing and castration to be easily performed on the animal.
(6) Grooming may be an essential stimulus in milking.

Grooming should be done regularly, as well as when considered necessary. Handling and taming should be done with caution and preferably by an experienced hand.

## 5.2 FEEDING FARM ANIMALS

### 5.21 ESSENTIAL CONSTITUENTS OF ANIMAL FOOD

Food is necessary for animals just as much as for human beings, to supply energy for their body and work and for body building. Therefore the food that animals eat must contain adequate amounts of nutrients for energy and body building. These essential nutrients are supplied from many sources and include water, carbohydrates, fats and oils, proteins, minerals and vitamins. The contribution of these food sources to animal nutrition is discussed below.

*Water*

Not less than 50 per cent of the animal body is made up of water, which is being constantly used and got rid of from the body through respiration, perspiration and urine. It has to be constantly replaced through drinking and as an essential constituent of animal foodstuffs.

Water in the animal body has a number of uses:

(1) All body fluids (e.g. blood, lymph) are water solutions.
(2) Secretions, hormones and digestive juices are effective in liquid form with water as the main solvent.
(3) Water is an essential transport medium for carrying food solutions (*ingestion*) and waste solutions (*excretions*) to and from every part of the animal.
(4) Water is necessary for the regulation of body temperature; this is particularly important in all parts of the hot West Africa.
(5) Water is necessary for the very many chemical reactions (e.g. hydrolysis) that go on continuously in the animal's body from day to day.

The water requirements of any particular animal will, however, depend on a number of factors, such as: (a) the type of animal (category, breed, age, size); whether the animal is in production or resting (e.g. a cow in milk, i.e. lactating, needs a lot of water); (c) environmental conditions, particularly in West Africa; and (d) the type of food being consumed (grass, succulent, or concentrates). Unless for special reasons (e.g. ill health or medical attention), the best management approach is to ensure that cool clean water is *always available* for the animal to take as required.

*Carbohydrates*

Next to water, carbohydrates constitute the largest portion of animal feedstuffs and the main source of energy. Carbohydrates (CHO) consist of carbon (C), hydrogen (H) and oxygen (O). There are different groups of carbohydrates based on complexity of the different elements. The simplest of them are the *monosaccharides* or simple sugars, as found in fruits and sweet potatoes; these are easily absorbed by the digestive systems of animals and man. The *disaccharides* are more complex sugars, as found in malt (*maltose*), cane or beet sugar (*sucrose*) and milk (*lactose*).

The third category of carbohydrates is the most complex group, including *starches, glycogen* (animal starch in liver and muscles) and *cellulose*. These three are referred to as the *polysaccharides*. Starches are found in a stable stored form in all cereals, tubers, fruits and seeds. The starches undergo digestion in animals, and the constituents are finally regrouped into glycogen. Cellulose, on the other hand, as in cotton, is found in all plants. Ruminant animals and termites are the only living organisms that can readily digest cellulose. Roughage and grasses contain a high proportion of cellulose and *lignin* in their plant cells and tissues, which when fed to ruminants are digested with the help of special bacteria. These polysaccharides get broken down into simpler carbohydrates in the animal's rumen, to release the energy in them for the animal's use.

*Fats and Oils*

Fats and oils, sometimes called *lipids*, both have virtually the same chemical composition, except that fats are solid and oils are liquid at room temperature. They contain carbon, hydrogen and oxygen in the form of esters of *glycol* and *fatty acids*. When consumed by animals and man, fats and oils are easily hydrolysed by the enzyme *lipase*.

Several tropical plants which are consumed by man and animals contain a significant amount of fats and oils (e.g. palm oil, groundnuts, melon seeds, coconuts, soya beans, sesame).

Body fat is stored in animals in a hump (e.g. camel, cattle), under the skin, around the tail (e.g. sheep), around the kidneys and in muscles. The tenderness of the meat of some temperate animals is a result of the thin fat layers deposited along the muscle fibres, which is not usually the case with animals in the tropical belt. The heat effect in the tropics discourages the deposition of fat in the animal body; hence only little quantities of fats and oils are necessary or used in animal diets in West Africa.

*Protein*

Animals use the protein constituent of some plants (e.g. legumes) and other animal tissues to build their own body. Protein is therefore essential for *all* animals: young, old, sick and nursing. Protein is required by animals for the production of enzymes, enzyme secretion and other body organ secretions in addition to being used in the repair of worn-out tissues.

The more common proteins in living tissues can be grouped into: (a) *simple* proteins (e.g. *albumins*), and (b) *conjugated* proteins (e.g. *chromoproteins, nucleoproteins*). All proteins are composite of *amino acids*; the conjugated proteins of course include other compounds or prosthetic groups.

The common foods can be grouped according to their protein content as rich, average or poor in protein, as shown in Table 5.2.

Table 5.2 *Protein in some common foods, in percentage by weight (i.e. grams per 100 g of food, or ounces per 100 oz).*

| *Rich* | | *Average* | | *Poor* | |
|---|---|---|---|---|---|
| Legumes: | | Wheat | 15 | Cassava | 1·5 |
| e.g. cowpeas | 22 | Maize | 10 | Potatoes | 2 |
| groundnuts | 27 | Rice | 8 | Fruits | <1·5 |
| soya bean seed | 40 | Sheep milk | 6 | Vegetables and fruits: | |
| Fish (dried) | 63 | | | e.g. carrot | 1 |
| Milk (dried) | 26 | | | pear | 1·5 |
| Meat (beef) | 19 | | | mango | <1 |
| Insect (locust) | 20 | | | coconut | 4 |
| Egg | 30 | | | | |

Different protein foods contain different combinations of amino acids. Generally, animal protein contains a more balanced proportion of the essential amino acids than protein from plants. Hence it is said that animal protein feeds have a higher biological value since most of their amino acids are used by the animals that eat them.

Protein requirements depend on several factors:

(1) *Age of animal:* Younger and growing animals need more than old ones.
(2) *State of production:* Pregnant and milking cows, nursing animals and laying hens need more than when they are dry or not producing.
(3) *State of health:* Sick animals need more protein as fortification against disease attack and during recuperation (recovery from sickness).
(4) *Type of animal:* Some animals, particularly the ruminants, are able to synthesise some types of amino acids in their stomach from

the non-protein nitrogen sources in the food they eat; hence they require relatively less protein in their diets, provided they have access to good quality pasture.

*Minerals*

Animals and man require many minerals for different body functions, as shown in Table 5.3. The principal minerals in alphabetical order are calcium, chlorine, cobalt, copper, iodine, iron, magnesium, manganese, phosphorus, sodium and zinc. These are required constantly by all animals.

*Vitamins*

Vitamins are very complex chemical compounds which all animals (including man) need, but only in small doses. They are essential for metabolism in all farm animals. Deficiency symptoms due to lack of vitamins are known as *vitaminosis* (a type of malnutrition). Generally, a balanced diet contains sufficient amounts except for vitamins D and K, which are rare in feeds and have to be supplied directly if deficient. Vitamin K is synthesised readily by intestinal bacteria, and vitamin D is easily manufactured in the skin of adult animals exposed to a reasonable amount of sunlight. Further discussion is given in Table 5.4.

*Classification of Animal Feedstuffs*

Food of all kinds fed to livestock are collectively known as *feedstuffs*. Those consumed by human beings, however, are referred to as *foodstuffs*. Feedstuffs can be classified on the basis of their food content and the proportion of fibre in the feedstuffs. Five common groups are known:

(1) *Roughages and succulents* include grasses, succulent herbage, hay, silage, and other green vegetables with high fibre and water content respectively. Most of these fibres (made up of cellulose and lignin) are not readily digestible by farm animals except ruminants (e.g. cattle, sheep, goats), which have a specially adapted alimentary canal for the digestion of such rough plant substances. Succulents, on the other hand, are usually green vegetables with little of the fibre content and with a high water content. Most animals and birds (poultry) can easily digest them, although in small quantities when compared with ruminants. Both roughages and succulents supply appreciable quantities of nutrients and vitamins in different amounts to the animals that feed on them.

(2) *Concentrates* contain a large amount of nutrient per unit of weight and volume when compared with any other class of feedstuff. Feedstuffs such as cakes (from groundnut, palm kernel, cotton seeds and soya bean), cereals, oil seeds and legume seeds supply

Table 5.3 *Minerals for animal use.*

| *Name* | *Major Supply Sources* | *Uses* | *Deficiency Symptoms* |
|---|---|---|---|
| Calcium and phosphorus | Bones, shells, fish, grasses, legumes, bone meal, fish meal. | Bone formation, skeleton, teeth. | Rickets; poor bone development; infertility; bone diseases. |
| Chlorine (sodium chloride) | Common salt, salt-lick, bone meal, vegetables. | Control of water metabolism; osmotic pressure of the cells; other vital body fluids (e.g. blood, lymph) and organs. | Reduced appetite; skin lesions; swollen lymph glands; drop in output of milk, eggs, etc. |
| Cobalt | Salt-lick, bone meal, vegetables. | Formation of vitamin $B_{12}$ synthesised by ruminants. | Blood disease (anaemia); emaciation. |
| Copper and iron | Salt-lick, seeds, fruits, colostrum. | Copper is catalyst for iron in formation of haemoglobin (in blood cells). | Anaemia; emaciation; loss of appetite; poor disease resistance. |
| Manganese | Seed coats, rice bran. | Present in all animal tissue; essential for enzyme formation; reproduction and lactation. | Sterility; perosis in chickens; poor tendon formation; poor conformation; low productivity. |
| Magnesium | Bone meal, mature grasses, some seeds. | Skeleton and bone formation; activates enzyme formation. | Low blood pressure; retarded growth; nervousness; tetany. |
| Iodine | Salt-lick, greens, colostrum. | Only in traces as thyroid hormone (thyroxin). | Enlargement of thyroid gland (goitre); retarded growth. |
| Zinc | Plants and animal tissues, milk, colostrum. | In all tissues for healthy functioning and food absorption. | Slow growth and poor development of hairs and coat colour. |

Table 5.4 *Vitamins for animal use.*

| *Name* | *Major Supply Sources* | *Uses* | *Deficiency Symptoms* |
|---|---|---|---|
| Vitamin A | Milk, eggs, cod liver oil, palm oil, animal fat, fruits and vegetables. | Normal growth; resistance to disease. | Loss of appetite; retarded growth; night-blindness; poor disease resistance. |
| Vitamin $B_1$ (thiamin) | Cereals husks, rice bran, yeast, liver, beans, peas, eggs. | Formation of body proteins (amino acids). | Beriberi; inflammation of nerves; paralysis. |
| Vitamin $B_2$ (riboflavin) | As $B_1$. | As $B_1$, especially for outer coats and body membranes. | Skin lesions; skin cracking. |
| Nicotinic acid (niacin) | As $B_1$. | Skin coat formation; nerve functioning. | Pellagra; skin diseases; diarrhoea; and poor nerve co-ordination. |
| Vitamin C (ascorbic acid) | Pawpaw, citrus fruits, guava, vegetables, grasses. | Outer membranes; body capillaries kept in good functioning order. | Scurvy; skin haemorrhages; swollen gums; teeth loss. |
| Vitamin D (calciferol) | Milk, liver, animal fat, fish. | Activates calcium and phosphorus absorption and synthesis in the body. | Poor bone formation; poor calcium and phosphorus absorption; rickets. |
| Vitamin E (tocopherol) | Cereals coats, green vegetables, milk, eggs, butter. | Reproduction; maintenance and absorption of vitamin A. | Sterility in animals; poor ovulation. |
| Vitamin K | Green vegetables, liver, and synthesised by intestinal bacterial action in animals. | Blood clotting. | Reduced ability for blood clotting; may lead to bleeding to death. |

protein, fats, oils and carbohydrates in appreciable quantities. They have low fibre content; hence they are used as the main diet source for non-ruminants and as supplementary feed in the case of ruminants. They are usually fed in granulated or milled form as mash.

(3) *Animal products* are the best source of balanced protein. They contain the essential amino acid derived from animal products, i.e. protein of high biological value. Hence, in livestock feeding, products such as fish meal, blood meal and milk (e.g. colostrum, whole milk, skim milk, ghee) are very useful feeds. They are used only in small quantities, partly because they are usually very expensive.

(4) *Mineral feed:* Under very intensive management, animals are not free to roam about and browse at plants and pick up a few of the essential minerals. They are stall-fed on prepared rations, most of which may not contain the essential minerals in required quantities. At such times, minerals in the form of common salt (sodium chloride), charcoal, shell or bone powder, salt-lick, etc. are supplied along with the food or separately in the pen.

(5) *Vitamin feed:* Usually animals will pick up a lot of their vitamin needs, but under intensive management or in special cases of illness vitamins may have to be supplied directly with the food to augment deficiencies.

## 5.22 ALIMENTARY CANAL AND DIGESTION

The mouth and the anus mark the two openings of the *alimentary system* or *canal* in all animals. The remaining organs concerned with the work of digestion and absorption are internal. The alimentary canal of every animal determines the type of food the animal can eat. On the basis of this there are three distinct groups of alimentary canals: that of non-ruminant animals (e.g. pigs, dogs, cats, rabbits); that of ruminant animals (e.g. cattle, sheep, goats); and that of birds (e.g. poultry). Figures 5.6 to 5.8 illustrate the digestive systems of a rabbit, cow and bird respectively.

The alimentary canal of each of these three groups performs five main activities with food:

(1) *Ingestion:* Taking in of the food usually through the mouth.
(2) *Digestion:* Breaking of the food particles into smaller bits for chemical and bacterial action in the alimentary canal.
(3) *Absorption:* Taking in of digested food in solution form into the bloodstream.
(4) *Assimilation:* Using of the absorbed food for energy, body building or repair of tissue as required.
(5) *Excretion:* Getting rid of body waste material and undigested food out of the alimentary canal.

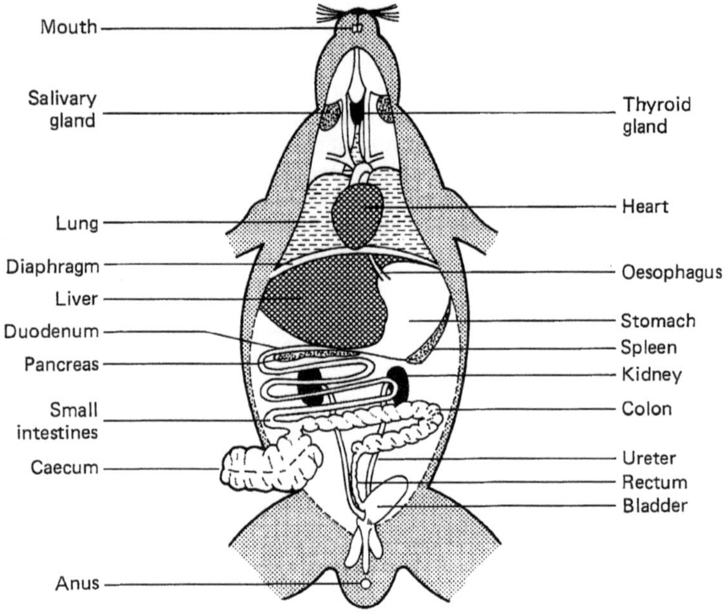


Figure 5.6 *Digestive system of a rabbit.*

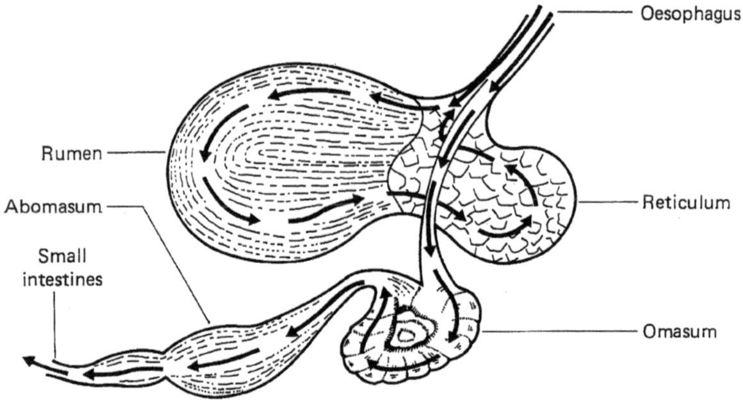


Figure 5.7 *The four sections of the stomach of a ruminant (cow).*

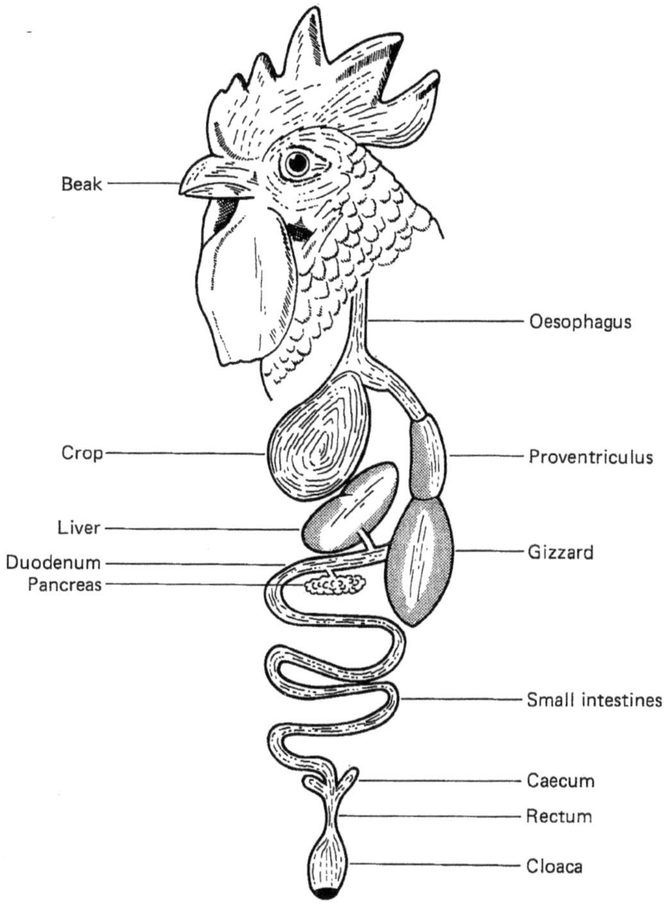


Figure 5.8 *Alimentary canal of a fowl.*

All three groups carry out the digestion in stages, which can be classified as being (a) *physical* and mechanical tearing and breakdown; (b) *chemical* reactions of enzymes on the food, leading to further breakdown; and (c) *bacterial* action on the food.

It is after the final digestion action that absorption takes place. To understand fully the process of digestion it is useful to follow the progress (as in Table 5.5) of the food through the alimentary canal of a simple non-ruminant such as the pig. The pre-digestion phase of

Table 5.5 *Digestion in a non-ruminant (e.g. pig, rabbit).*

| | *Nature of Action* | | |
|---|---|---|---|
| *Area of Action* | *Physical and Mechanical* | *Chemical* | *Bacterial (Biological)* |
| Mouth | Mouth is opened; food is engulfed (taken in) and chewed with the help of the teeth, tongue and cheeks; then food is swallowed. | Saliva is mixed with the food; enzymes of the saliva start the digestion of carbohydrates. | None |
| Oesophagus | Contractile muscle movements of the canal (peristalsis) pass food down to the stomach. | Mixing of saliva continues. | None |
| Stomach | Bowel muscle movements mix food and juice; peristalsis pushes food down to the duodenum. | Gastric juice is secreted containing enzymes (pepsin and renin) and hydrochloric acid (HCl at 0·2–0·5N), which begin digestion of proteins. | None |
| Duodenum | Peristalsis continues. | – Liver produces bile stored in the gall bladder; bile emulsifies fat and oils into drops of colloids.<br>– Pancreas secretes pancreatic juice which has three enzymes: amylase for sugars; lipase for fat hydrolysis; and proteinase for converting protein into amino acids. | None |

Table 5.5—*continued.*

| *Area of Action* | *Nature of Action* — *Physical and Mechanical* | *Chemical* | *Bacterial (Biological)* |
|---|---|---|---|
| | | – Duodenal wall cells produce enzymes too, acting on carbo-hydrates, proteins, fats and oils. | |
| Small intestine | Peristalsis continues. | All actions that started in the duodenum continue through the length of the small intestine. | None |
| Large intestine | | | |
| (a) Caecum | Peristalsis continues. | Absorption of water and digested foods. | Some bacterial action on food remnants, mainly on polysaccharides (cellulose, lignin). |
| (b) Colon | Peristalsis continues. | | Decay through bacterial action on indigestible food and waste materials. |
| (c) Rectum | Waste expelled through anus. | | Bacterial decay continues; wastes expelled. |

identifying the food by smelling or sight (by odour in birds) is not considered, even though this must be remembered. Tables 5.6 and 5.7 follow the digestive process in a ruminant (cattle) and bird respectively.

### 5.23 PRINCIPLES OF NUTRITION

Nutrition covers the adequate supply of food needs of livestock. The previous discussion examined the type of foods that animals eat and the

Table 5.6 *Digestion in a ruminant (e.g. cattle).*

| *Area of Action* | *Nature of Action*<br>*Physical and Mechanical* | *Chemical* | *Bacterial (Biological)* |
|---|---|---|---|
| Mouth | As in non-ruminants. | As in non-ruminants. | None |
| Oesophagus | Peristalsis pushes food down into rumen. | As in non-ruminants. | None |
| Stomach | | | |
| (a) Rumen (in young animals, e.g. calf, rumen is not fully developed, say until 6–8 weeks) | Peristalsis continues; peristalsis then pushes food from rumen into reticulum. | Absorption of fatty acids and ammonia into the bloodstream; vitamins B and K synthesised are absorbed. | Micro-organisms break down fibres and produce acids, fatty acids and ammonia in gaseous forms; micro-organisms synthesise vitamins B and K. |
| (b) Reticulum | – Acts as a sieve separating big food particles from smaller ones.<br>– By anti-peristalsis of oesophagus, cud-like food is then forced out of reticulum back to the mouth for rechewing ('chewing the cud').<br>– Chewed paste then forced through oesophagus into omasum. | Very little | Very little |

Table 5.6—*continued.*

| *Area of Action* | *Nature of Action* | | |
|---|---|---|---|
| | *Physical and Mechanical* | *Chemical* | *Bacterial (Biological)* |
| (c) Omasum | Peristalsis is very strong, churning food paste; food is then moved into abomasum. | Water is absorbed. | Very little |
| (d) Abomasum | As in stomach of non-ruminants. | As in stomach of non-ruminants. | Very little |
| Small intestine | As in non-ruminants. | As in non-ruminants. | Very little |
| Large intestine | | | |
| (a) Caecum | As in non-ruminants. | Some absorption of water and digested foods (cellulose etc.). | (a) Heavy digestion of cellulose, lignin and other polysaccharides by bacterial action. |
| (b) Colon | | | (bc) As in non-ruminants. |
| (c) Rectum | | | |

way the food is digested and utilised by the different types of animals. Now we have to examine the most effective way of supplying the food needs of farm animals.

When animals are fed on a type of food that contains all the essential food items (ingredients) in the proportion required by the animals, then the food is called *balanced diet*. The quantity of that food or balanced diet supplied per day is called the *ration*.

When the ration is a balanced diet and in such quantity to meet only the essential needs of the animal without consideration for other production or for work, the animal is said to be on *maintenance ration*; i.e. the ration is supplied only for the animal's life processes, such as respiration, digestion, circulation and excretion. However, when the animal is used for work, or is producing meat, milk or eggs, the animal's food need is considerably increased, and the food supply to be adequate

Table 5.7 *Digestion in a bird (e.g. poultry).*

| *Area of Action* | *Nature of Action* — *Physical and Mechanical* | *Chemical* | *Biological (Bacterial)* |
|---|---|---|---|
| Mouth | Birds have no teeth, so no chewing; pick up food with beaks and tongue and swallow directly. | Very little | None |
| Oesophagus | | | |
| (a) Thin oesophagus | Passage of food by peristalsis. | Very little | None |
| (b) Crop (dilated part of oesophagus) | Storage of food; mixing and softening. | Juices from glands in the crop walls soften food. | None |
| Stomach | | | |
| (a) Proventriculus (upper part) | Peristalsis continues. | (ab) Gastric juice and similar reactions to stomach in non-ruminants. | (ab) Very little |
| (b) Gizzard (lower portion; very muscular) | Peristalsis continues; small sharp stones in gizzard have grinding effect. | | |
| Duodenum, Small intestine, Large intestine | As in non-ruminants. | As in non-ruminants. | As in non-ruminants. |
| Rectum | Opens into the cloaca, which performs dual function of excretion of wastes from the intestine and from the bladder (urine) in the case of a bird; the cloaca is also used as the genital opening in birds. | | |

must allow for these extras. When the ration supplied is such that takes all these into consideration, it is called *production ration.*

Furthermore, in special cases (e.g. pregnant cows) special rations are prepared to help the special needs of the animal. The foetus has to be fed and the milk organs (the udders) that will provide the food for the young ones later have to be adequately reinforced if the animal is not to suffer from *malnutrition.* Just before delivery, pregnant cows are given special additional ration. This is known as *steaming-up ration.* The extra ration is mainly protein; it helps to induce increased activity of the animal and foetus.

Malnutrition can come about from several sources, such as inadequate *quality of food.* Food fed may be lacking in the required calories (energy) and/or protein (digestible nitrogen) and/or minerals or vitamins. These deficiencies can be corrected by supplying the required amounts of each category of food. Quality of food is also affected by excess of one category of food at the expense of the other. More calories than the amount required by the animal will lead to deposits of extra fat in the animal body. This is the greatest nutritional problem in livestock feeding in West Africa. Owing to the high price of protein supplements, vitamins and mineral foods, livestock keepers try to economise with disastrous results by trying to substitute protein with carbohydrates.

Inadequate *quantity of food* is another common source of malnutrition. In West Africa, the nomadic existence of livestock men and their livestock results from the search for green grasses and water during 5–6 months (October–April) of the dry season. To avoid this constant (perennial) problem of dry-season food shortages, the feeding of supplementary food in the form of hay, silage and concentrates in the dry season is very important. If dry season feeds are not supplied, the animals lose weight and condition. They also become less resistant to disease and pest attack.

*Feeding Young Animals*

Young animals (e.g. calves, piglets, chickens) have very small, simple and weak digestive systems. Nutrition at the early part of their lives determines not only their survival but also their growth pattern for the rest of their lives. At this particular time also, they are at their weakest (most vulnerable) in their effort to resist disease and pest attacks.

For all these reasons, feeding them is an important and delicate matter. The essential requirements in feeding young animals include:

(1) *Colostrum:* This is the yellowish white milk produced immediately after parturition; it is rich in vitamins, minerals and disease-resistant organisms and secretions collectively known as *antibodies.*

(2) *High-level proteins, minerals and vitamins in diets.*
(3) *Hygiene and sanitation in feeding:* The food should be free from undesirable impurities and fed to the young animal at a temperature as close to body temperature as possible.

## 5.24 PASTURE IN ANIMAL FEEDING

A pasture is defined as where animals graze for food. In section 4.3 some pages were devoted to the discussion of grasses and cover crops. The student's attention is drawn to this section again. In West Africa, there are large areas of natural pastures in the savanna belt. In addition, there are established pastures which are specifically cultivated just like any other crop.

All ruminants (e.g. cattle, goats, sheep) have the advantage of being able to convert roughages (e.g. grass, legumes, pastures, either green or in the form of silage or hay) into milk, meat and energy for work or traction. The quality of product and the ability of these ruminants to survive on pastures depend on three essential factors: (a) quality (composition and age) of the herbage in the pasture; (b) carrying capacity and animal population (or stock rate); and (c) maintenance of the pasture. These points can now be discussed briefly.

### *Quality*

The quality of the herbage is important because the digestion of protein into amino acids will depend on the available proteins in the herbage; similarly with minerals and vitamins. Quality of pasture depends on the constituents of the pasture: the type and proportion of grass or grass–legume mixture, its growth, taste, and how much of it the animal eats.

Quality of pasture is also influenced by the age of the herbage. When plants are too young and succulent, the water content is high and animals may eat too much of them, thereby filling their stomach with too much cellulose and less of other nutrients. On the other hand, if the herbage is too old, it contains a lot of fibres and less of protein; the animal feeding on such pasture will not get enough protein. The best pasture is mature (not old) grass.

The quality of the pasture is improved by keeping a balanced proportion of good grasses and legumes mixtures, and by avoiding grazing animals on either too young grasses/legumes or on too old herbage.

### *Carrying capacity*

The number of animals of any particular type that can be conveniently fed on a hectare (or acre) of pasture is known as the *carrying capacity* of the pasture. Carrying capacity is based mainly on the quantity of available herbage and on a quality factor with respect to composition and age of pasture, as discussed above. Furthermore, the carrying capacity varies in respect of whether the pasture is a natural grassland or established pasture. With established pasture, the *biological value* (i.e. the amount of animal product possible from the given unit of pasture) can be easily determined from the composition of the pasture.

Whenever the number of animals on any pasture land exceeds the carrying capacity of the pasture, the animals are likely to be underfed. In addition, too many animals on a pasture land may lead to overcrowding, overgrazing, poorer pasture, soil erosion and a rapid spread of pests and diseases among the animals.

*Maintenance of Pasture*

A few essential points in keeping the quality and quantity of food from any pasture are:

(1) Keep weeds off pastures.
(2) Regular grazing or cutting is essential to avoid too old herbage.
(3) Avoid overgrazing and overcrowding.
(4) Supply water (irrigation) and fertiliser to pasture as necessary.
(5) Practise rotational grazing.
(6) When pasture is poor, do zero grazing and supplement with concentrate feeds.

*Making Hay and Silage*

The provision of dry season feed is one of the most important problems in raising livestock in West Africa. Feeding of concentrate supplement is one effective way of solving this problem. As grasses and legumes do not thrive in the dry season, it is useful to store grasses for the dry season in the form of hay or silage.

In making *hay*, the grass/legume pasture is cut at the appropriate maturity; not too old and not too young, preferably just before flowering, to give the maximum possible amount of food nutrients. The cut pasture is left in the open for a day or two and later arranged in stacks on a tripod and allowed to dry. It is then bundled and kept in the hay store until needed. Drying should be gradual so that moisture is removed slowly without necessarily destroying most of the food nutrients in the cut pasture.

*Silage* is preserved without drying the fodder. It can be prepared and stored in any one of three systems: (a) the stack silo, (b) tower silo (above ground), or (c) pit silo.

Grasses, or grass–legume mixtures, or any other form of cereal or succulent herbage can be used in making silage. The material to be used is cut at full maturity: not old, just before flowering. The cut material is chopped into 5–7 cm (2–2·5 in) pieces, which are then placed in the appropriate container. Here the chopped material is compressed by pressure (feet, pounders, tractor weights, etc.) to reduce oxygen and thereby prevent rotting (putrefaction). When a layer as thick as 0·3 m (1 ft) is formed after compression, another load/batch (consignment) of chopped material is added and compressed to another height or depth of 0·3 m (1 ft), until either the pit is full or the available material is all compressed. Then an earth mound 0·3 m (1 ft) thick is made over the silage. This is to keep off rain, sunlight and air. The silage is not to be opened until it is required for use. Once opened, it has to be used. The activities of cutting, chopping, compressing and making the mound cover should be done when there is no rain and should not last more than 2 days. If they take longer, some decay may set in and poor silage is produced.

## 5.25 FEEDING FOR PROFIT

The first reason why farmers keep cattle is to sell them for cash. Therefore feeding farm animals must be done in a profitable way, i.e. in an economical manner. If the farmer wants to make profit, four important points to bear in mind are:

(1) Select the best animal(s) for the product in mind.
(2) Feed adequately and regularly.
(3) Use as much as possible of cheap local feedstuffs.
(4) Sell the animals at the appropriate weight and age the market can take; otherwise they become too fat, overweight and not of prime quality.

### *Select the Best Animals*

The ability of farm animals to convert pasture feed concentrates and farm wastes into meat, milk and eggs depends on two important relevant factors: the *breed* and the *feeding*. For each type of animal, different breeds have different capabilities of food conversion. Table 5.8 gives different capabilities for different animal categories.

Once the best breed is selected for the particular output the farmer has in mind, the next important thing is to give the animal the most balanced food in adequate amounts. The breeds included in Table 5.8 are only those that have survived under local conditions of heat, high humidity and disease resistance.

### *Feed Adequately and Regularly*

Adequate feeding means feeding the right quantity and quality and at regular intervals. Feeding too much in the hope that the animal may

Table 5.8 *Animal breed outputs in West Africa.*

| *Type of Animal* | *Name of Best Breed for* | | | |
|---|---|---|---|---|
| | *Milk* | *Meat* | *Work* | *Eggs* |
| Cattle | Friesian<br>Muturus<br>Zebus | Muturus<br>N'damas<br>Zebus<br>Hereford | Zebus | — |
| Pigs | — | Large White<br>Saddle Back<br>Large Black | — | — |
| Sheep | — | Local breeds | — | — |
| Goats | — | Local breeds | — | — |
| Poultry | — | Rhode Island Red<br>Austrolops<br>Local breeds | — | Leghorn<br>Rhode Island Red |

grow faster not only is a waste of food and money but also may do harm to the animal. Quantities to feed and the proportion of nutrients to be supplied can be scientifically calculated, based on age, weight and type of animal. It is good to seek advice on this from specialists and government officials.

Regular feeding keeps the animals' systems and organs healthy, and growth is improved. Food conversion into meat, milk or eggs by animals is at a maximum when the animals' systems are functioning very well.

*Use Cheap Local Feeds*

There are a lot of locally produced feedstuffs, fruits and vegetables which often go to waste but which can be put to good use in feeding farm animals. It is true that the food nutrient contents of most of these feeds are not known. Nevertheless, they can be used to a limited extent from time to time. Many of the local fruit-factory wastes and farm wastes (e.g. peelings, haulms) can be put to good use in livestock feeding without any serious problems.

When feeds such as concentrates are purchased, care should be taken to avoid losses from rats and other rodents, and waste of any type should be reduced to a minimum. Feeds should be mixed in the correct proportion so that the gain in weight is as expected.

*Sell at the Right Time*

Poor feed users in the herds should be *culled*, i.e. removed from the herd and sold or slaughtered. In this respect the keeping of yield records of all animals is very important and useful.

When the animals have reached their market size, age and weight, i.e. are mature enough for sale, they should be disposed of. Keeping animals that are too mature not only wastes food but also encourages these animals to grow fat and lose condition. They therefore become less than prime for the market. Their prices drop because their quality is lower. It is always advisable to plan ahead and prepare the animals for market when prices will be good in the market. This has the advantage of reducing the costs of production on the farm and at the same time increasing the prices the farmer gets; his net income or profit is thus higher. Selling the animals before they are mature is equally bad, as they are likely to bring less cash income and less profit.

## 5.3 REPRODUCTION IN FARM ANIMALS

### 5.31 GENERAL PRINCIPLES OF REPRODUCTION

Reproduction, or the natural act by which the younger ones (offspring) are produced, is an essential part of livestock husbandry. A knowledge of reproduction is very important if livestock is to be kept at a profit and if the output of livestock products is to be improved. It is through the act of reproduction that good qualities, as well as bad ones unfortunately, are passed from parents to offspring. This process of transferring qualities from parent to offspring is known as *inheritance*. The parts of the cell responsible for inheritance transfer are the *chromosomes*. Every living cell has chromosomes, which are usually in pairs and are usually referred to as *homologous chromosomes*.

There are two types of cells in the animal body: the *somatic cells* and the reproductive cells (*gametes*). In every animal the somatic cells contain equal numbers of homologous chromosomes (in pairs), and the number of such chromosomes is said to be *diploid*. But in the reproductive cells the chromosome pairs have been split between two separate cells during genetic cell division (meiosis). The number of chromosomes on each gamete is thus half that in the somatic cells and is called the *haploid* number.

During *mating*, the male gametes (*sperms*) are introduced into the female reproductive organs. After mating, the sperms swim towards the female gametes (the *ova* or eggs) and fuse with them. This fusing is called *fertilisation*. Successful fertilisation results in the formation of a new cell (the *zygote*). The zygote develops into a tissue known first as the *embryo* and later as the *foetus*, and finally emerges from the female parent (mother) as the offspring.

How often a young one is produced depends on the type of animal, state of health of the animal, fertility of the male and female animals at mating, and the *fecundity* of the mother, i.e. the ability of the mother to carry and develop the embryo into a foetus successfully. For breeding purposes, all these features must be taken into serious consideration.

The nomenclature (names) of farm animals vary with sex and age. Some common ones are given in Table 5.9.

## 5.32 MORPHOLOGY OF REPRODUCTION ORGANS

Every farm animal can be identified as male or female by the external and internal features of its reproductive organs. Even in birds (poultry) where the sexual organs are not obvious, all being internal, some external features are indicative. For instance, the wattles, comb and plumage in the male are generally more lustrous, colourful and highly exaggerated compared to those of the female bird; the male bird is also larger than the female.

The male reproductive organs (see Figure 5.9) consist of the *scrotum*, a bag-like organ in which is encased usually two *testes*. The testes produce the sperms (male gametes), which are then transferred through

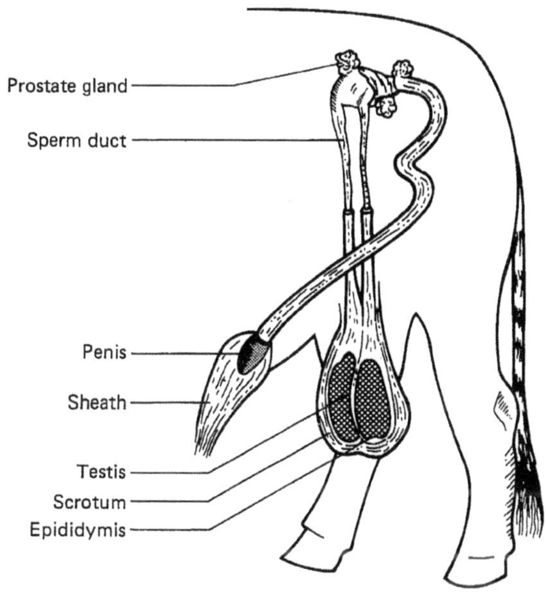


Figure 5.9 *Reproductive organs of a bull.*

Table 5.9 *Names (nomenclature) of some farm animals for the different sexes and age groups.*

| *Animal Type* | *Adult*[a] | | | | *Young* | | *Baby* |
|---|---|---|---|---|---|---|---|
| | *Age* | *Male* | *Female* | *Castrated Male* | *Male* | *Female* | |
| Cattle | 2 years | Bull | Cow | Bullock | Young bull | Heifer | Calf |
| Pig | 6 months | Boar | Sow | Hog | Young boar | Gilt | Piglet |
| Sheep | 6 months | Ram | Ewe | Wether | Tup | Kid ewe | Lamb |
| Goat | 6 months | Billy goat | Nanny goat | Castrated goat | Kid goatling | Kid goatling | Kid |
| Poultry | 5 months | Cock | Hen | Capon | Cockerel | Pullet | Chick |
| Horse | 2 years | Stallion | Mare | Gelding | Colt | Filly | Foal |
| Dog | 6 months | Dog | Bitch | Castrated dog | Puppy | Puppy | Puppy |

[a] The adult female name is given immediately at first pregnancy. In the case of hens, it is given at the first laying of eggs. With sterile females, age is often used.

the *sperm duct* to the *penis* and out. The penis contains spongy tissues which when filled with blood becomes considerably hard, elongated and enlarged. In this state of erection, the penis is easily introduced during mating into the female sexual organ before the sperms are deposited into the female receptacle.

The female reproductive organs (see Figure 5.10) are in two parts. The sexual organs for producing the female gametes, receiving the male gametes and developing the embryo constitute one set. The mammary glands for nourishing the young offspring constitute another set. The male animal also possesses mammary organs, but these are rather vestigial and not used for nourishing the young and are therefore not developed.

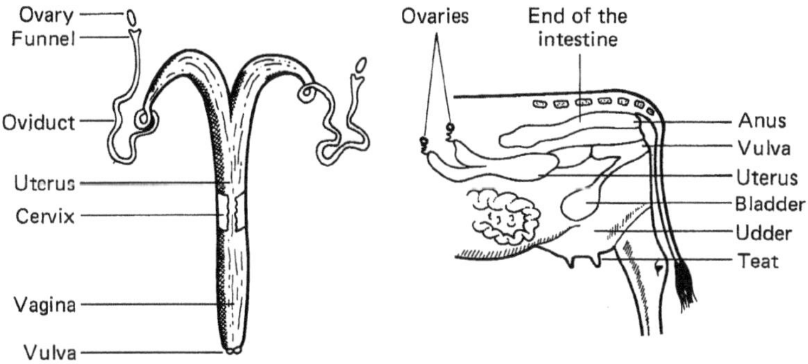


Figure 5.10 *Reproductive organs of a cow.*

The main sexual organ of the female animal, the *uterus*, opens out externally as the vaginal opening or *vulva* just below the anus. An egg or ovum from the *ovary* wall gets detached and drops into the *funnel* receptacle. From there it is pushed along the *oviduct* where it is fertilised by an oncoming male gamete. If there is no male gamete there is no fertilisation, and the egg aborts and is washed out of the uterus through the *vagina*. On the other hand, if the egg is fertilised a zygote is formed, and this is attached by the *placenta* to the lining of the uterus where the embryo is finally developed into a foetus.

In poultry (see Figure 5.11) the principle is the same, even though the physiology is slightly different. The *cloaca* serves as the opening of the genital, urinary and rectal outlet. The production of gametes is similar, and after mating the male sperm is deposited and fertilises the female egg. If there is no sperm there is no fertilisation, but in this case the egg does not abort. However, it remains sterile, and the egg produced will never hatch into a chicken. The female gamete, whether fertilised

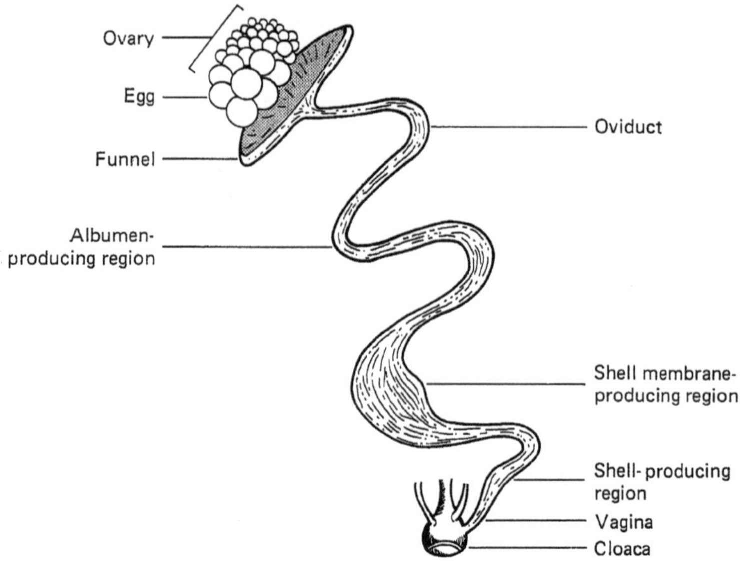


Figure 5.11 *Female reproductive organs of a bird.*

or not, moves into a food reserve of *yolk* and *albumen*. This food reserve, and the female gamete or zygote (if fertilised), are subsequently encased in a calcareous *shell* and then ejected (pushed out) through the cloaca of the hen. Whereas with mammals the embryo develops into a foetus inside the female (in the uterus), with birds the embryo develops into a foetus outside the female. Hence the protective shell cover of the egg is needed in all birds, and reptiles (see Figure 5.12).

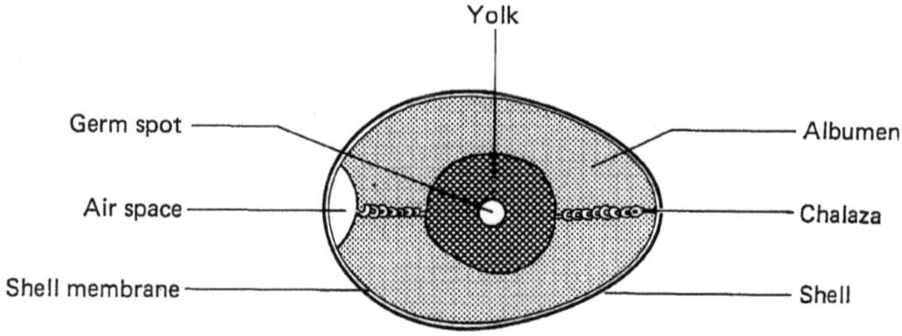


Figure 5.12 *An egg opened horizontally.*

## 5.33 FERTILITY AND CARE OF PREGNANT ANIMALS

*Fertility* in farm animals refers to the ability of the male to produce and deliver healthy (i.e. virile) sperms that will always fertilise the egg. In the female it refers to the ability of the female to produce and deliver fertile eggs that will be readily fertilised by the sperm. In common language, fertility will also cover the animals concerned, i.e. the male being always willing and in good form to mate with the female. On the part of the female, the term will be used to include regularity of heat, showing the obvious signs of heat, and readiness to take the male. The breed (genetic factors), food and health of the animals are the essential factors influencing fertility.

*Fecundity* is another similar term. It means the ability of the female to develop healthy embryos and foetuses regularly. This attribute is essentially a genetic factor. Offspring from fecund parents tend to be also fecund. In pigs, for instance, the presence of a large number of teats on the sow indicates her ability to nurse a given number of piglets adequately. In selecting a breeding sow it is therefore highly essential to count the number of teats and to select only sows with not less than ten teats for breeding.

With birds (poultry) high fecundity is indicated by the number of eggs laid within a year. An average of 200 eggs a year is usually considered a reasonably good record.

### *Heat Periods and Mating*

The female animal will not take the male, i.e. will not allow the male to mate with her, unless and until she is on *heat*. A female is therefore said to be on heat when she is showing signs of wanting the male. These heat signs usually coincide with the ovulation cycle, and the maximum heat is attained when the animal is most likely to be fertilised.

Heat signs in the female animal include restlessness, usually with mating calls and cues. The vagina is swollen with blood, is red and exudes slimy juice. The female sniffs around other animals and tries to mount them. If a mature male is around, through his sense of smell he too gets excited and his penis is extended and enlarged. When female animals are pregnant, heat period ceases. *Mating* is the act of putting the female to the male and the completing of introducing the sperm into the vagina.

### *Artificial Insemination*

This is the act of artificially introducing sperm into the vagina. It can be done successfully only when the female animal is at the peak of her heat. The sperm (or the *semen*) is collected from the male using a dummy female or 'teaser', and by masturbating on this the male is induced to release his sperm into a specially designed tube (or artificial vagina). The sperm is stored at a very cold temperature until required. At insemination the sperm is heated to the animal's body temperature and introduced if and when necessary into the female, using this time an artificial penis.

Artificial insemination is practised when the male animal to be used is a genetically superior (or *proven*) male. Semen is easier and cheaper to transport in its diluted and frozen state than is an animal (e.g. a bull). Therefore, in breeding work where the male is far away, or too expensive to be kept by any one farmer, artificial insemination is practised extensively so that the good qualities of the proven male animal is spread over a very large number of female animals. It is a successful method of improving livestock, but it requires careful handlers for successful operation.

*Pregnant Animals*

Once the egg is fertilised, the zygote develops into an embryo in the uterus. The embryo becomes the foetus with a placenta for conducting nutrition from the mother to the foetus and for removing wastes and excretion from the foetus to the mother. Pregnant animals usually need more care, especially at the latter part of pregnancy.

Food is the most important item during pregnancy. The female animal now has to be supplied enough food for both herself and the foetus. In the last 2 weeks of pregnancy, *steaming-up ration* is essential to maximise development, ease delivery and increase milk output after delivery. Pregnant animals have to be housed, preferably in individual pens, and supplied regularly with protein-rich foods, vitamins, minerals and a lot of water. Regular simple exercise is necessary daily, and the animal should be kept free from flies and other pests and disease sources.

At delivery (*parturition*) the foetus is pushed out normally with either the head between the two forefeet or the two legs first. But if there should be any difficult parturition, a veterinary officer should be called as soon as possible. Delay is always dangerous. Immediately after parturition the mother starts to produce milk. In the first few days after parturition a special type of milk called *colostrum* is produced. It is yellowish white in colour and very rich in protein (12–15 per cent), and it contains a relatively high proportion of antibodies, minerals and vitamins which are extremely beneficial to the baby animal. Discussion on the care of baby animals will be continued in Chapter 6 for the individual classes of animals.

Table 5.10 gives a useful summary of the reproductive cycle of some farm animals.

Table 5.10 *Useful biological information and terminology.*

| *Female Animal* | *Heat Period (Every)* | *Pregnancy (Gestation Period)* | *Parturition (Delivery) Terminology* | *Frequency of Parturition (Average)* | *Weaning at* |
|---|---|---|---|---|---|
| Cow | 21 days | $9\frac{1}{2}$ months | Calving | 3 times in 4 years | 6 months |
| Sow | 21 days | 100–115 days[a] | Farrowing | 2 every 9 months | 8 weeks |
| Ewe | 21 days | 150 days | Lambing | 3 times in 2 years | 3–5 months |
| Nanny goat | 21 days | 150 days | Farrowing | 3 times in 2 years | 3–5 months |
| Hen | Every day[b] | Egg incubation: 21 days | Hatching | Every 21 days[b] | 3–4 weeks |

[a] Easily remembered as 3 months, 3 weeks, 3 days (+3 days occasionally).
[b] Except when self-incubating eggs (21 days) and nursing chicks for 3–4 weeks.

## 5.4 IMPROVING FARM ANIMALS

### 5.41 AIMS AND METHODS OF BREEDING

The objective of any livestock keeper is to produce good quality animals as a means of producing good quality products. Hence the science and art of breeding good farm animals is the backbone of any livestock raising. The two essential factors determining good farm animals are *genetic factors* and *environmental factors*, such as the climate, management and food under which the animals are raised. *Breeding* is the science of keeping and improving the *genetic potential* or capability of the animals.

It was mentioned in section 5.31 that animals reproduce by the fertilisation of the female sex cells (ova) by the male sex cells (sperms). These sex cells contain haploid chromosomes, through which the characters and potentials of the parents are passed on and inherited by the offspring. The idea in breeding is to identify animals with the desirable potentials and to encourage such animals to reproduce more and more of their kind. For instance, the quality of good milk yield is passed on from mother to daughter; such characteristics are known as *hereditary characteristics*.

The important aims of breeding are:

(1) To increase the potential capacity of the animals to produce more eggs, milk or meat or, in the case of work-bulls, to do more work.
(2) To improve the quality of such products, such as better yolk, better milk, better quality of meat and better temperament (mood) to work better or carry heavier loads.
(3) To improve and increase the tolerance or resistance to the prevalent diseases (e.g. trypanosomiasis, pleuro-pneumonia, rinderpest, foot and mouth) that afflict animals in West Africa.
(4) To improve the tolerance of the animal to the environment, including heat, heavy rainfall, temperature, etc.
(5) To evolve an animal most suitable for a particular purpose (e.g. raising horses for playing polo has a different aim from raising horses for show jumping; shepherd dogs are bred purposely different from hunting dogs).

There are several breeding methods, but the two simple and most common ones are: (a) by *selection* from existing animals; and (b) by *breeding* (inbreeding, cross-breeding and upgrading or introduction). These methods are discussed in the next section.

### 5.42 SELECTION AND BREEDING

*Selection*

One of the oldest method of improving livestock is by selection from the existing group of animals. In carrying out selection, the animals possessing and exhibiting the desired (required) hereditary characters (or *traits*) are used more and more for mating and for producing offspring. This is done because these characters pass on from mothers to daughters, and from mothers through the male also to their daughters.

The advantages of the selection method are:

(1) It is the cheapest method since it does not involve buying new stock
(2) The ability of the animal to be tolerant or resistant to local and prevalent diseases is maintained.
(3) The type and level of feeding and management are not necessarily changed; hence local conditions are not adverse for the animal.
(4) It is easy to undertake.

There are some disadvantages too:

(1) It is a very slow process, especially if the animals are not very fertile or fecund. For instance, with cattle it takes at least three consistently good lactations to select a good cow, and the bull from that cow a year after will mature 3 years later, to give a total of 4 years (including the lactation). After selecting the bull, we allow one more plus three, giving 8–9 years to have the calf from the selected bull, making a total of 9–10 years. By this time the bull is too old for good mating performance.
(2) Certain undesirable (unwholesome) characters are difficult, if not impossible, to remove or eliminate.
(3) The potential of the herd is limited by the potential of the best animal in the herd, which may not be much compared to that of other animals not being selected; hence it is of limited scope.
(4) In any selection method there is the danger of inbreeding.

*Breeding*

Improvement of farm animals through breeding is a very common scientific practice. There are three possible alternative breeding methods: (a) upgrading or introduction, (b) cross-breeding, and (c) inbreeding or pure line breeding.

*Upgrading* means introducing new blood from another animal with better qualities. Indigenous livestock in most parts of West Africa have some desirable qualities (e.g. tolerance to some local diseases and pests and to heat, adaptation to poor forage and low level of maintenance).

For these reasons they provide, in their own poor way, some form of income and food for peasants. Their performance can be easily and cheaply improved in a significant way by upgrading. Proven bulls of better animal types have been used extensively, either directly or by artificial insemination, to upgrade such cattle. The better male (*sire*) is used to mate a female (*dam*) of low grade stock; the first offspring is half of sire and half of dam. If this offspring is again mated to the better sire (*back-crossing*), the next offspring is three-quarters of the sire. After six such crosses, the offspring will possess 98 per cent of the good sire's character.

This method is cheap and made even cheaper through the introduction of artificial insemination. The disadvantage in this method is the slowness of the improvement. With cattle it will take over 6 years to get to the 98 per cent level of the good sire. Furthermore, as more and more of the sire is introduced (progressively), the other good characters of the original dam are weakened or lost.

*Cross-breeding* is done when different breeds are mated. While this is going on the two breeds are also being maintained pure and separately. The two breeds to be used must separately possess some qualities that are desired in the offspring. The first offspring is called a *hybrid* or $F_1$ progeny (first filial generation) and is supposed to possess a higher standard than either of the parents.

The merits of this method are: (a) *hybrid vigour*, and new combination of chromosomes; (b) formation of new genera in offspring (increase *heterosis*); and (c) speed, since the hybrid is produced within a year. The disadvantages include the cost of maintaining two pure breeds of animal. The hybrid will not breed true to type; i.e. their offspring may behave more like either parent or combine the worse traits in both parents. Finally, the process of cross-breeding is highly scientific and requires the keeping of a number of records; hence in West Africa it is limited in practice to special breeding centres, research stations and universities.

*Inbreeding* is done when close relatives are mated to produce offspring. This is done to increase the *homozygosity* (similar genes) in the same animal. Continuous inbreeding leads to the creation of a *pure line*.

The advantages are the cheapness, the ease of operation and the advantage of concentrated traits or genes of a particular type. The disadvantages include the obvious problem of having undesirable characters also being concentrated. It is risky if there is a new type of epidemic, which may lead to total loss. Finally, animals so pure-bred are difficult to establish in any place other than where they were first made into a pure line breed.

## 5.5 DISEASES AND PESTS OF FARM ANIMALS

### 5.51 BASIC ANIMAL HYGIENE

Animal hygiene is essential on the farm for a number of reasons:

(1) Animals in good health grow and reproduce faster.
(2) Products such as milk, meat and eggs are produced in greater quantity, have better quality and carry no risk to the human beings who use them.
(3) Diseases when they occur are expensive to cure or eradicate (cost of drugs, and other losses).
(4) Losses in the breeding herd can lead to a lapse (delay) of several years in a breeding programme or to complete loss of all breeding animals.

Hygiene on the farm is important with respect to:

(1) *The environment:* Housing and grazing areas have to be clean.
(2) *Feeding:* Stale and dirty foods are breeding areas for disease-causing organisms.
(3) *Grooming:* Regular cleaning of animals helps to keep them clean, warm, docile and productive.
(4) *Injuries:* These should be quickly and constantly attended to, since open wounds attract flies and other pests and may expose the animal to secondary infections.
(5) *Attendants:* These should be clean and not become carriers of pests and diseases from one animal to another.
(6) *New stocks:* If not quarantined and carefully examined, new stock may lead to a new source of infection to all the herd.

Furthermore, there are four categories of animals that for one reason or another need constant attention: (a) breeding animals, (b) pregnant and nursing animals, (c) young animals and (d) sick animals. These should *always* be given the best attention possible at all times. In the case of the breeding herd, the two aspects of cost of replacement and the transmission of diseases and pests through contact make them the most important group of animals in the herd. Usually the male of the herd is worth half the whole herd, and at one time or another he is in contact, during grazing and at mating, with the female members of the herd. The other categories (the pregnant and nursing animals, the young and the sick) as a group are the weakest and are most likely to be attacked and suffer seriously. Therefore hygiene on the farm should concentrate very seriously on this group of animals.

Depending on the nature and degree of infection, diseases in farm animals may be divided into four classes:

(1) *Epizootic* disease is prevalent among all animals temporarily (e.g. tick-borne diseases).
(2) *Endemic* disease is prevalent among animals in a particular area regularly (e.g. trypanosomiasis).
(3) *Infectious* disease is easily transmitted by air, water or food (e.g. anthrax).
(4) *Contagious* disease is transmitted by contact (e.g. virus diseases).

As a general rule, all unknown disease symptoms should be reported to the veterinary officer for action.

Usually in farm animals, sickness or ill health are easily recognised in a number of ways, including: (a) laziness in moving; (b) poor or loss of appetite; (c) salivation at the mouth; (d) lacrymation (i.e. water in the eyes); (e) lesions in the mouth; (f) animal becomes temperamental and avoids being touched; (g) tendency to isolation; (h) erratic and slow movement of animal; (i) swellings on joints which are painful to touch; and (j) loss of coat colour (not shiny and thick).

### 5.52 DISEASE-CAUSING ORGANISMS

Table 5.11 gives a simple categorisation of disease-causing organisms and the nature of the diseases they cause. Tables 5.12 to 5.14 give an analysis for each category, the nature of infection, and the transmission of some of these diseases.

Table 5.11 *Important disease-causing organisms.*

| *Viral* | *Rickettsia* | *Bacterial* | *Protozoan* |
|---|---|---|---|
| Foot and mouth<br>Fowl pox | Heart water | Anthrax<br>Black quarter | Fowl coccidiosis<br>Red water |
| Rinderpest | | Contagious abortion | Trypanosomiasis |
| Swine fever | | Foot rot<br>Fowl cholera<br>Fowl typhoid<br>Mastitis<br>Pneumonia and pleuro-pneumonia<br>Tuberculosis | |

Table 5.12 *Viral diseases.*

| *Name* | *Transmission* | *Animals Usually Affected* | *Symptoms* | *Treatment and Control* |
|---|---|---|---|---|
| Foot and mouth | Direct contact; food, and other materials in contact with infected animal | Cattle, pigs, sheep, goats | Sores on tongue, lips, feet, heads; increased salivation; limping walk; laziness; death | Vaccination; isolate infected animals; killing if epizootic |
| Fowl pox | As above | Poultry | Swellings (wart-like) and sores on combs, wattles and mouth parts | As above |
| Rinderpest | As above | Cattle | Ulcers in mouth; high fever; excreta blood stained; diarrhoea | As above |
| Swine fever | As above; also through ticks | Pigs | As above with rinderpest in cattle | As above; it is cheaper to kill infected animal |

Table 5.13 *Bacterial diseases.*

| *Name* | *Transmission* | *Animals Usually Affected* | *Symptoms* | *Treatment and Control* |
|---|---|---|---|---|
| Anthrax | Open wound; food/water already contaminated with the bacteria | All farm animals except poultry | Very dangerous and lethal, causing death without symptoms; high fever | Vaccination; antibiotics too; bury or burn dead animals |
| Black quarter | As above | Cattle, sheep | Lameness; swollen hips, joints, shoulders; high fever; and death | As above |
| Contagious abortion | As above | Cattle, sheep, pigs | Abortion after 7 months of pregnancy | Vaccination; don't breed from infected animal |
| Foot rot | As above; through bacteria in soil and pen | Sheep, goats, donkeys | Lameness; sore hoofs | Use antibiotics on sores; clean out pen; always use dry bedding |
| Fowl cholera | As above for anthrax; through direct contact | Poultry | Weak, dull-looking, drowsy birds; sudden death | Vaccination; antibiotics; bury dead birds |

Table 5.13—*continued.*

| *Name* | *Transmission* | *Animals Usually Affected* | *Symptoms* | *Treatment and Control* |
|---|---|---|---|---|
| Fowl Typhoid | As above | Poultry | High fever; loss of appetite; diarrhoea; death | Vaccination; destroy infected birds and infected pens |
| Mastitis (disease of the udder) | Through the teat canal or wounds on udder | All *female* livestock except poultry | Hard shrunken udder; blood-stained milk | Antibiotics; infected animals to be milked by someone not milking other cows; isolate |
| Pneumonia or pleuro-pneumonia | Inhalation of bacteria; contact and food | All farm animals | Cough; rapid breathing; shivering; fever; loss of appetite | Isolate; vaccination |
| Tuberculosis | As above | All farm animals | Coughing; discharges from mouth and nostrils; diarrhoea; loss of colour; death | Isolate; seek veterinary assistance; cheaper to kill animal infected |

Table 5.14 *Protozoan diseases.*

| *Name* | *Transmission* | *Animals Usually Affected* | *Symptoms* | *Treatment and Control* |
|---|---|---|---|---|
| Fowl coccidiosis | Stale feeds; contaminated food and water | Poultry | Bloodstained diarrhoea; anaemia; sudden death of chicks | Use sulpha-drugs to inhibit protozoa |
| Red water | Ticks | Cattle, sheep, goats | Stiff walking; urine red with blood; loss of appetite | Remove ticks; dipping; antibiotics and sulpha-drugs |
| Trypanosomiasis | Tsetse flies; wild animals as secondary hosts | All farm animals (not important in poultry) | High fever; emaciation; loss of appetite; swollen limbs; death | Eradicate tsetse; clear forest of undergrowth and wild animals; sulpha-drugs and some prophylactics. Use resistant animals; male-sterile flies |

*Viral diseases* are caused by viruses, which are ultra-microscopic organisms. Their effect is usually devastating (excessive) because the diseases are easily transmitted by contact, and by the time the disease condition is identified or has become noticeable the virus has affected most parts of the animal body. Viral diseases tend to become epizootic (see Table 5.12).

*Rickettsia diseases* are caused by small micro-organisms which in size are midway between viruses and bacteria. There are not many of them, and their important disease activities are limited. The important one in West Africa is the *heart water* disease in ruminants, which is transmitted by ticks. The disease is characterised by high fever, body muscle convulsions, uncontrolled movements of the body, and death. The only control is by deticking, dipping and destroying seriously infected animals.

*Bacterial diseases* are caused by bacteria, which constitute the most versatile group of disease-causing organisms. These micro-organisms are not visible to the naked eye, but are seen only through a microscope. Bacterial diseases are all readily controlled through strict hygiene measures which prevent the survival and spread of bacteria. Besides, a number of vaccines have been produced. These can effectively control bacterial diseases and should be given to the animal as a matter of routine. Prevention is better than cure (see Table 5.13).

*Protozoan diseases* are caused by protozoa, which are single-celled organisms visible not to the naked eye but only with a microscope. There are not many, but two very serious protozoan diseases of livestock in West Africa, trypanosomiasis, and coccidiosis (in poultry), have accounted for a great number of deaths in farm animals for a number of decades. All protozoan diseases are accompanied by high fever, loss of appetite, emaciation of the animal and death within 2 weeks. All protozoan diseases are easily controlled by the use of sulpha-drugs (see Table 5.14).

### 5.53 DISEASE MANAGEMENT

The important point to remember in disease management is that prevention is better than cure; it is also cheaper. There are five main means of preventing diseases.

*Environment*
Hygiene, good feeding, clean environment and the use of only healthy animals for breeding are the essential points. In addition, diseases can be prevented on farms through:

*Immunity*
Some animals have natural immunity to some diseases, or are relatively tolerant, and can resist or overcome mild doses of infection. This is only

effective of course if the animal is in good health, is well fed and is kept in clean surroundings. Immunity can also be given to the animal through the use of *vaccines* and immune *serums*.

Natural immunity can be: (a) inherited from the parents; (b) actively acquired after surviving an infection; or (c) passively acquired through the placenta or the colostrum of the mother.

*Control of Ticks and Pests*

A full discussion on pests (e.g. ticks, worms, insects) follows in section 5.54 but ticks and other pests not only harm farm animals but also transmit diseases. Any control and prevention of diseases must include the control of the population of ticks and other pests. This control is largely achieved through a number of ways, including:

(1) Avoid overcrowding of animals.
(2) Overgrazing pastures help to keep a high tick population on pastures.
(3) Remove ticks from animals; *burn* all ticks removed.
(4) Routine hand dressing, spraying or dipping for *all* farm animals; dust poultry with insecticides. Chemical compounds usually used in tick control are arsenic compounds, chlorinated hydrocarbons, or organic phosphorus compounds. All these chemicals are also poisonous to human beings and should be handled with care.
(5) Against flies, fly-proof pen for sick animals, and pedigree stock is advisable; hanging of fly catchers in pens is helpful; darkening of fatteners' pen reduces fly activities.
(6) Worms and other intestinal parasites are readily controlled through regular deworming and good feeding.

*Vaccination*

Vaccination is used as general control measure against a number of animal diseases. Vaccinations in West Africa have been developed and used effectively against a number of animal diseases (e.g. anthrax, pleuro-pneumonia, rinderpest, swine fever). With poultry, diseases such as fowl cholera, fowl typhoid, Newcastle disease and fowl pox are easily avoided through routine vaccination. Serum inoculation is also frequently used.

*Control of Animal Movement*

This is a desirable way of checking the spread of disease and is practised in most parts of West Africa. Animals are also given regular vaccination and inoculations at control posts where the movement of animals from place to place is put under control.

*Quarantine* is a method whereby new animals imported into a country, or into a new herd, are kept under inspection away from the others. During the quarantine period, any infection that is likely to be in incubation at the time of their arrival is given time to break out and treated before the animals are allowed to join the crowd. In addition, quarantine is a useful period for the new stock to become acclimatised (i.e. get used to the area, climate and food).

*Isolation* of sick animals immediately illness is noticed gives the herdsmen not only the time to prevent the spread of disease, but also an opportunity to give the sick animal the treatment and full attention necessary for curing it.

*Use of Chemicals*

Chemicals such as antibiotics, sulpha-drugs and disinfectants are useful medications to have around and to use on the farm in the battle against diseases. It is important to note that chemicals are expensive and should not be wasted. They are also poisonous and should not be used freely without medical advice; otherwise they may be used in a manner that may hurt rather than help the farm animals.

To summarise, a few useful hints in the prevention and control of diseases are:

(1) Good hygiene in the farm, farm buildings, environment, food, water and attendant is highly desirable and profitable.
(2) Good feeding strengthens the animal against attack by diseases and pests.
(3) Avoid contact with infected animals; quarantine new herd or stock as of routine.
(4) Chemicals (e.g. dipping mixtures, disinfectants, antibiotics, sulpha-drugs, dehorning drugs, vaccines, inoculation serums) are useful in the fight against diseases and pests; they should be made ready on farms and used with the advice of veterinary officers.
(5) Report all sick cases at the government veterinary centre before too late.

## 5.54 MAJOR PESTS AND PARASITES OF FARM ANIMALS

*Pests* are living things that depend for their food on their hosts (i.e. animals). They are also called *parasites*. They do not directly cause diseases, but when they transmit diseases they are called *vectors*. The major pests of farm animals in West Africa can be grouped into two categories. Those that live within the tissue of the animal hosts are *endoparasites*, while those that live on the outside of the animals are *ectoparasites*. Those that feed on other things (e.g. house fly) but still have a nuisance effect on farm animals are *simple pests* and not parasites.

Parasites feed on digested food (e.g. worms) or the blood of animals, giving rise to loss of condition, poor growth and sometimes death (e.g. termites). Others are vectors of fatal diseases; the tsetse fly, for instance, carries the fatal disease trypanosomiasis and the tick carries deadly diseases such as red water. Many other pests, such as flies (e.g. maggot flies), are attracted by open wounds and lay their eggs on the wound; their larvae hatch and may become parasitic, feeding on the open tissues of the animals. Others (e.g. house fly) can be simply irritating to the animals, but they may also carry bacterial infection by contact from one animal to the other or to the food and water meant for animals.

The major parasites and pests of farm animals in West Africa are categorised in Figure 5.13.

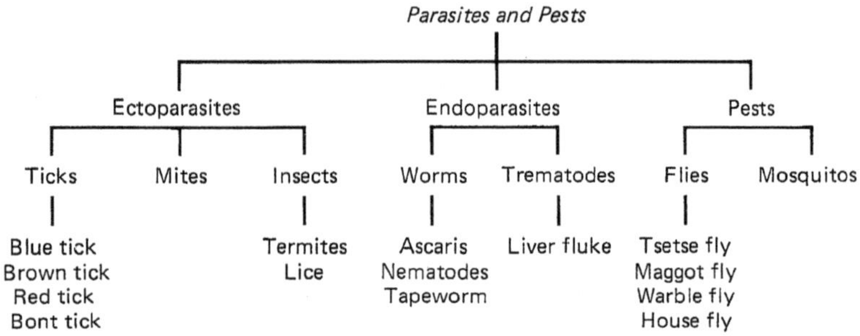


Figure 5.13 *Major parasites and pests of farm animals.*

### *Ticks*

Ticks feed on animal blood, leading to loss of blood, emaciation and even death in some farm animals. They are particularly dangerous because as vectors they transmit diseases. The blue, brown and red ticks transmit red water disease. They also cause irritation when they bite farm animals, thereby making the animals restless and uncomfortable.

The knowledge of the lifecycle of any particular parasite helps in the control as well as cure of the parasite and hence of the disease it transmits. There are four stages in the life (*metamorphosis*) of a tick: the *egg* hatches into a *larva* with six legs; the larva develops into a *nymph* with eight legs; and the nymph develops into an *adult* with eight legs. This is represented in Figure 5.14. The larvae, nymphs and adults develop in the ground. After each phase change, the new phase climbs on to a plant and waits for the host animal. On the host animal, it sucks

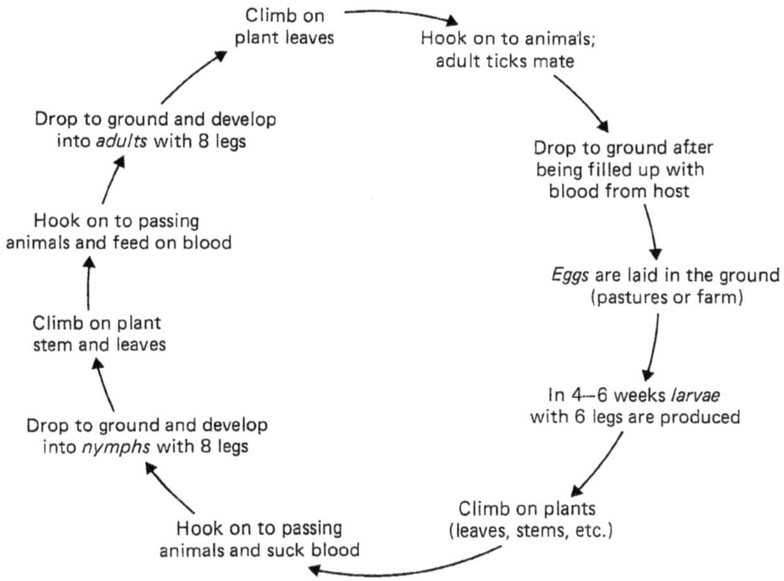


Figure 5.14 *Lifecycle of a tick.*

blood and drops to the ground, then develops into another phase and again gets on to a plant.

Tick control is by picking and burning, or by spraying or dipping with some compounds (e.g. arsenic), also by avoiding overgrazing. Regular grooming and general cleanliness reduce ticks on farm animals. The practice of rotational grazing helps to reduce their chances of completing their life cycle, hence reduces infestation.

*Mites*

Mites are ectoparasites which feed on the blood of poultry, usually at night. Mites can eat under the skin too, causing loss of feathers, drop in egg yield, emaciation and sometimes death.

*Insects*

*Termites* are ectoparasties too, and they bite and cause irritation. In large numbers they may cause death. Control is by dusting fowls with insecticide and disinfectants. Roosts and poultry houses should be surrounded by gutters filled with water sprayed with oil and/or disinfectant, to prevent termites crawling to the animals at night. The areas around the poultry pen should be cleared of bushes and shrubs, and termite hills on farms should be destroyed.

*Lice* suck the blood of all farm animals. Like ticks they usually hide at the juncture of ears and legs and between hoofs. In poultry they hide at the base of the feather. In cattle, sheep and goats they also hide at the base of hairs. When infestation is heavy the blood sucking can lead to loss of condition, emaciation and death, especially in young poultry. As with ticks, they are controlled by picking, dipping, spraying and hand treatment with poisonous chemicals and burning. Birds can be dusted and their pens fumigated or sprayed with disinfectants.

*Worms*

Worms are parasites living in the organs and tissues of farm animals and feeding on their digested food. They cause irritation, loss of food, emaciation and, if too many, constipation, constriction of the alimentary canal and sometimes death. Three important worm parasites of farm animals in West Africa are the *ascaris*, *nematode* worms and the *tapeworm*. Control of worms is by dosing with antihelminth worm expellers.

Ascaris worms are very common in the small intestine of pigs and sheep. *Ascaris suum* is the most common one. Apart from feeding, the young worm can migrate to the liver and lungs of animals and damage these organs. Most people avoid eating pigs because of *A. suum* worms.

Nematode (round) worms are in all animals, where they feed on the digested food of the animals. Some (e.g. hookworms) suck the blood of the host, giving rise to anaemia, emaciation, loss of weight and diarrhoea.

Tapeworms (*Taenia solium* or *T. saginata*) are very harmful to livestock and man, causing loss of food, anaemia and emaciation. The tapeworm has two hosts, with man as the primary host and cattle or pig as the secondary host. *T. solium* uses the pig and *T. saginata* the cattle as the secondary host.

The segments of the tapeworm are each called a *proglottis*, which when filled with eggs become detached from the other parts to start a new life. The *egg* passes out of man's faeces and later develops within it an *embryo* with six hooks, which may be swallowed by a pig or cattle and taken into the alimentary canal. From there it travels into any muscle and forms a *bladder worm* or *cysticercus*. This bladder worm if eaten by man develops into another tapeworm in man's intestine, where it develops proglottides and the cycle continues. Figure 5.15 illustrates the tapeworm lifecycle diagrammatically.

Liver Flukes (*Fasciole hepatica*)

The liver fluke is an endoparasite, feeding on farm animals' food and destroying quite a few organs, especially the liver. The sheep is usually its most common victim, making the animal weak, emaciated, anaemic from serious bleeding and occasionally causing death. The snail is the

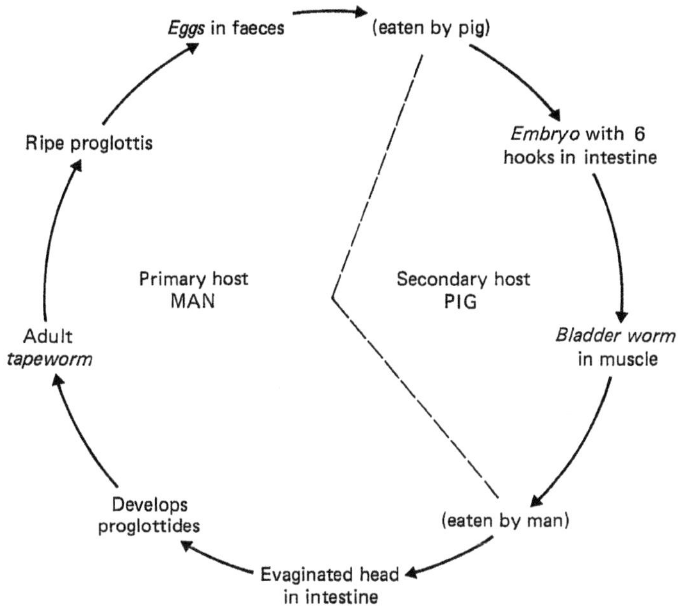


Figure 5.15 *Lifecycle of a tapeworm.*

secondary host, thus making an effective control of the disease very difficult indeed. The life history of the liver fluke is given in Figure 5.16.

The liver fluke, a hermaphrodite flat worm (trematode), has a leaf-like body which moves through the action of its muscular suckers and burrows its way through tissues, causing much damage. From the lifecycle, it can be seen that one control method is by destroying the snail host. The other method is to keep the sheep away from the metacercaria cyst during grazing.

*Trypanosomiasis and the Tsetse Fly*

The most deadly disease of livestock, particularly cattle and pigs, is trypanosomiasis, caused by *Trypanosoma* spp., a protozoan that spends its life in three hosts: cattle, the tsetse fly and wild animals. The wild animal is immuned to it and hence serves as a carrier, thus making control very difficult. The protozoan can continue its lifecycle between the cattle and the tsetse fly (*Glossina* spp.), of which there are many types found in different parts of tropical West Africa.

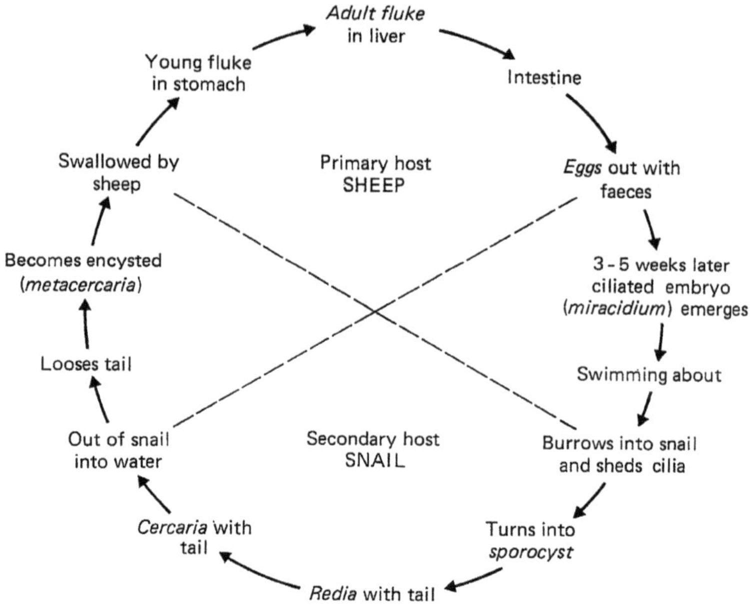


Figure 5.16 *Lifecycle of the liver fluke.*

The protozoa are transmitted by two methods. It can be spread through direct contact, when a fly (with sucking mouth parts, e.g. stomoxys, etc.) feeds on the blood of an infected wild animal or cattle and, carrying the protozoa on its mouth parts (the proboscis), immediately after goes to feed and suck the blood of a healthy animal; the protozoa are introduced mechanically. However, the more common method of transmission is when the tsetse fly sucks the blood of an infected animal and the protozoa develop in the fly and finally move into the salivary gland of the fly; during feeding the fly releases saliva into the animal tissue before sucking the blood of another animal, and the protozoa are introduced with the saliva into the body of a healthy animal. The cycle is repeated indefinitely. Sheep, goats, pigs and horses are also affected.

Control is by use of resistant animals (e.g. Muturus and N'damas and their crosses). The use of prophylactics helps to reduce (or attenuate) the effect of the attack by reducing the activity (virulence) of the protozoa. Curative drugs have not been very successful. Clearing of bush reduces both the fly and wild animal populations, thus reducing the risk of infection. But this is very expensive, and most poor countries of West Africa cannot afford the expenses involved. Recently the control of tsetse flies using a male-sterile technique to reduce their population has also been tried. Keeping animals healthy through good feeding and preventive measures is essential for survival against even the mildest attack.

*Chapter 6*

# Notes on Some Important Livestock of West Africa

A short simple classification of farm animals will be useful in enabling us to identify the important types of farm animals, poultry and fish in West Africa. Based on their feeding habits, the four-footed animals or mammals to be discussed in this chapter are divided into three broad groups as shown in Figure 6.1.

## 6.1 RUMINANT HERBIVORES

This class of farm animals (e.g. cattle, sheep, goats) have one thing in common. They all feed mainly on pastures, from which they derive the bulk of their nutritional needs in respect of energy, proteins, minerals and vitamins. This they can do because of the special nature of their digestive system, as discussed in section 5.22. The development and

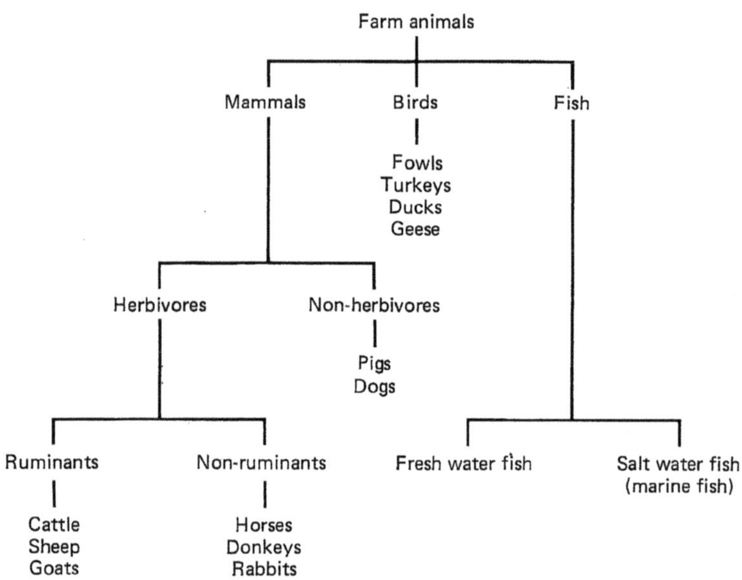


Figure 6.1 *Common farm animals in West Africa.*

management problems facing herdsmen and farmers raising this class of livestock in West Africa are three: (a) breeding stock, (b) feeds (including pasture), and (c) trypanosomiasis and other diseases.

*Breeding stock*

The existing breeding stocks for most of these animals as they exist in West Africa are not such as can induce confidence in intensive management practices, because the stocks are poor in fertility and poor in the production of meat, milk or energy for work. They are also basically poor converters of food; hence their productivity is low. This explains why in most West African countries, most government efforts in the area of livestock improvements concentrate on the selection and cross-breeding of animals with better breeding, better food conversion and higher productivity of meat, milk and energy.

*Feeds*

It is true that natural pastures abound in several parts of West Africa. Unfortunately, most of these natural pastures are of varying composition, quantity and quality. It is therefore difficult to determine their *carrying capacity* (i.e. ability to sustain a given number of animals). Furthermore, most of these natural pastures provide no herbage in the dry season, which sometimes lasts for 6 months. Animals that feed exclusively on these natural pastures therefore feed on pastures of unknown, but clearly poor quality pasture for only 6 months and starve for the remaining 6 months. This situation in part explains the nomadic/transhumance nature of livestock raising in West Africa.

The alternative to dry season starvation is to supply livestock with concentrate supplement, hay and silage in the dry season. But this is an expensive proposition which is clearly beyond the income or management level of most of the stockmen in West Africa. It is also important that where such costs have to be made the output from the stock in terms of meat, milk and energy for work must be more than enough to pay for such expenses. It is doubtful, as explained above, whether the breed of livestock now kept in West Africa can reach such a high level of productivity to pay for these extra costs.

*Trypanosomiasis*

This deadly disease of livestock in West Africa is a very serious limiting factor to livestock production in the area. Until this disease can be brought under control in a most inexpensive way, livestock raising will still be limited to the free zone area where the activities of the tsetse fly (carrier of the disease-causing protozoan) are not hazardous (dangerous).

There are other major widespread diseases (e.g. anthrax, foot and mouth, rinderpest, tuberculosis, pleuro-pneumonia) which also have to be put under control. Some of these diseases (e.g. anthrax, tuberculosis, foot and mouth) have to be completely eradicated because they are deadly to both livestock and man. The export potential in West African countries of livestock and livestock products cannot be encouraged until these deadly diseases are put under control or eliminated.

One area of livestock development for West Africa where great progress can be made lies in the integrated livestock/mixed-farming system, whereby livestock men are settled and cultivate pasture (improved natural and/or established pasture). Similarly, crop husbandry men (traditional farmers) are also induced to raise livestock. It is not easy, but it provides the only effective system in the long run for bringing about improved stock, the supply of adequate and good quality feed all year round, as well as effective control of trypanosomiasis and other deadly diseases.

## 6.11 CATTLE

### *Types and Breeds of Cattle*

From recorded history, cattle in West Africa originate from a series of nomadic migrations from western Asia through Egypt. The earliest ones are the *N'damas*, followed by the humpless short-horns (*Bos brachyceros*). *Muturus* cattle now predominate in the humid equatorial belt of West Africa. After them came the *Zebus* (*Bos indicus*). In West Africa today, the cattle population has mixed up a lot in breed even though several types can still be identified. These are indicated in Figure 6.2.

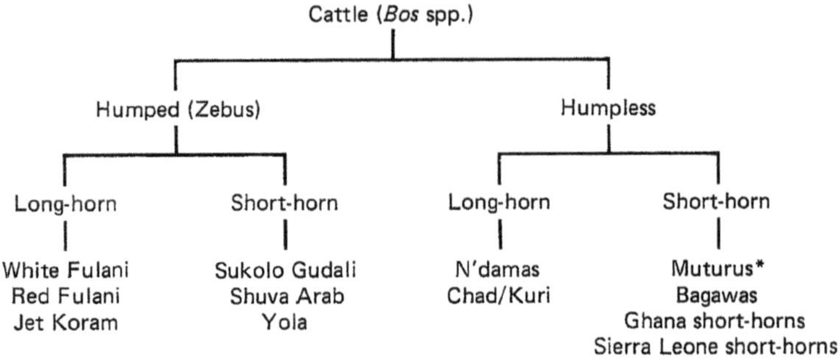


* The Muturus are collectively known as the West African short-horns.

Figure 6.2 *Cattle breeds in West Africa.*

On the basis of their products and configuration, cattle can be one of three types: beef cattle (for meat), dairy cattle (for milk) and work cattle (for power and traction). West African cattle types fall into none of these categories, even though compared to one another some are better than others for any particular product. The N'damas and the Muturus and the West African shorthorns in general are very poor for milk but much better for meat; on the other hand, the Zebus are better milk producers and good for work. Hence, breeding work in most West African countries is concentrating on making these animals better producers along the lines mentioned above, while keeping in the main stock their tolerance to local diseases, particularly trypanosomiasis. The

Table 6.1 *Essential features of good cattle.*

| *Beef* | *Dairy* | *Work* |
|---|---|---|
| *Body* heavy with thick muscles and long broad back. | *Body* slender; low body weight. | *Body* very heavy with heavy shoulders. |
| *Hump* heavy and solid with square shape. | *Udders* large and well developed. | *Back* long and broad. |
| *Legs* short and filled with flesh. | *Legs* average. | *Legs* long and sturdy. |
| Regular breeding; fast growing; early maturing. | Regular breeding; high fertility with regular calving. | Regular breeding; good frame. |
| Minimal fat deposits on body; mainly muscles, clean and heavy. | Milk yield heavy and at regular intervals with high butterfat content. | Chest strong and healthy; able to withstand drought. |
| Tolerant to local diseases, especially trypanosomiasis. | As with beef | Resistant to local diseases, especially trypanosomiasis. |
| Tolerant to heat. | As with beef | Highly tolerant to heat and high temperatures. |
| *Local types:* Muturus, N'damas, Zebus, Chad. | *Local types:* Zebus, Chad. | *Local types:* Chad, Zebus. |
| *Exotic types* (in West Africa): Friesian, Ayrshire, Jersey. | *Exotic types* (in West Africa): Hereford, German brown, Swiss brown, Boran cattle. | *Exotic types* (in West Africa): None; Asian buffaloes are now being tried. |

N'dama is particularly a good breed for meat and is reputed under good management and feed to be resistant to trypanosomiasis.

The essential features of good cattle for any of the three areas of production are given in Table 6.1. Horses are being introduced for work much more than cattle. The Zebu is probably the best local breed all round for meat, milk and work, except that it is not resistant to trypanosomiasis and hence is not for extensive use in the coastal areas of West Africa.

*Dairy Cattle Management*

The most important aspect of livestock raising is for milk and meat production. Milk is a very important and complete food, containing high proportions of protein (3·5 per cent), minerals (0·5 per cent), fat (3·5–4 per cent) and sugar (4·5–5 per cent). If the dairy cow is to produce these useful food items for human consumption and sales, it is important that the animal be well fed and well looked after. Doing these is the aim of cattle management. The quantity and quality of milk produced is influenced by four important factors, all of which are related to management: (a) age of the cow; (b) quality of food fed; (c) breed of the animal; and (d) state of health of the animal. Good management will therefore aim at selecting the best breed, keeping only mature, productive and not too old bulls and cows for breeding, feeding the best quality food in adequate amounts, and keeping the animals in good health through proper housing, a clean environment, regular vaccination, routine grooming, dipping for ticks and urgent veterinary attention as required.

*Calf Rearing*

The calf of today is the breeding stock of tomorrow. Therefore any attention and expenditure on the calf is an indirect investment which is likely to yield great dividends in the future. There are three aspects of calf rearing worth noting: feeding, housing and health.

*Calf feeding* is in three delicate stages: (a) the first month, (b) weaning, and (c) first grazing.

*The first month* is very crucial from the day of birth. An important point to remember during this period is that the calf's stomach is delicate and *cannot* digest any solid food – certainly not grass or roughages. The second point is that the milk from the mother is good balanced food containing appropriate amounts of nutrients for the calf. Besides, in the first week after calving the cow releases *colostrum*, a specially rich milk which contains hormones and antibodies to help immune and strengthen the calf against infection and to stimulate its rapid growth and development. The calf should suckle this or, if hand fed, it should be supplied in adequate amounts, at regular intervals and as close to the body temperature of the calf as possible. Fresh warm milk should be fed until weaning at the age of 6–9 months.

*Weaning* is another critical period. Before reaching weaning age the calf is on its own, trying to nibble at foods supplied to the mother. Where the calf is not with the mother, the nibbling may be on the 'beddings'. To avoid infection, the calf should be supplied with little amounts of protein-rich concentrates to nibble at from the age of 4 months, after which this is gradually increased while milk is still being supplied in gradually reduced amounts. At weaning, the calf is completely off milk and on solid foods.

*First grazing* is important in two respects. First, the intake of roughages by the calf will now be at the discretion of the calf. Therefore the quantity and quality of herbage consumed by the calf is regulated by making sure that the pasture for the calf is good and free from hard tough grasses that are too old. Grazing time should also be limited to avoid constipation, and it should not be allowed too early in the morning to minimise infection and cold from too wet grasses. Secondly, the first grazing marks the beginning of other sources of infection. Careful attention should be paid to ticks, flies and other sources of infection. The calf should not be overexposed to cold, rain or excessive heat, and preferably it should not graze with adults for too long, if at all.

*Housing* is important in calf rearing. Warmth, fresh air and clean bedding are the three important relevant points to bear in mind. Housing does not have to be elaborate, so long as it is not draughty nor exposed to rain and the bedding is changed regularly. Immediately after the birth the calf stays with the mother, which licks the calf for the first 2 hours. This should be encouraged as it cleans the calf's body (grooming) and keeps it warm.

The *health* of the calf is essential for good growth and survival. The common calf diseases are usually digestive problems, worms, cold and pest attack, in addition to other local diseases that attack other animals as well. Calf health will remain sound if proper precautions against infection with feeding, housing and clean environments are strictly followed. When there is any trouble, the veterinary officer should be consulted for advice. Operations such as castration and numbering are usually done after the weaning phase and are part of farm routine.

On the whole, cattle, whether for milk, beef or work, have to be properly fed, housed and kept in good health. All the essential points involved have been mentioned and should be carefully followed.

## 6.12 SHEEP

Sheep are kept in West Africa mainly for mutton (meat from sheep), with sheepskin of course as a useful by-product. The temperature is too hot for wool production, and the breeds are poor in milk.

The three important breeds of local sheep found in West Africa are: (a) Dwarf sheep (Pagan sheep); (b) Long-legged sheep, especially of the Uda breed, which is a religious meat animal for Muslims; and (c) the Macuna wool sheep, mainly of the Sudan and Ethiopia but found in the mountainous area of the Cameroons as well.

Sheep in West Africa are all kept under the free range and nomadic systems of management. They roam around, picking up food and pasture wherever and whenever they can find it. Kitchen wastes (e.g. tuber peelings) and farm residues (e.g. groundnut haulms, leaves, poor quality grains) are sometimes fed.

Sheep, like any other ruminants, need adequate feeding and housing if they are to produce good meat. Therefore the essential principles of good husbandry – feeding, housing and health care – should also apply to them.

### 6.13 GOATS

Goats, like sheep, are also kept in West Africa mainly for meat. Goatskin is a useful by-product for the production of fine leather. Goats are of course browsing animals, and in their efforts to eat anything around they tend to damage farm crops as well. The keeping of goats near farms is sometimes not recommended unless the animals are tied up (tethered), or the farms are fenced.

There are three breeds of goats that can be identified in West Africa: (a) the Dwarf goat of the Wet Equatorial (Coastal) Belt; (b) the Long-legged goat of the Sahelian zone, which has a better milk production than the Dwarf breed; and (c) the Sokoto (in Nigeria) Red Goat, which probably originated from Guinea but has over the centuries been well known for its good skin.

Goats, like sheep and cattle as a whole, to give their best need to be properly fed, housed and kept healthy.

## 6.2 NON-RUMINANT HERBIVORES

*Donkeys and rabbits* are useful non-ruminants that feed on grasses and herbages too. They do not have the complicated stomach pattern of the ruminants and as such cannot survive on only grasses. They need seeds, feeds and concentrates. They do not form part of any elaborate commercial livestock keeping in West Africa. They both survive on the free range system.

Rabbits are of course very small animals, but they reproduce fast – every 3 months – and as such they have been considered as a useful source of protein for the poorer classes of the West African population. However, they have yet to be kept on any extensive scale. Not much study has been carried out on them in West Africa. Because they are cheap and small, school children can conveniently look after them, both as pets and for commercial purposes.

Housing rabbits is essential if they are not to destroy crops and/or run away. Cages are used, but mud houses can also be built, provided the floor is made of concrete strong enough to prevent the rabbit from burrowing holes through it and escaping. Feeding consists mainly of nuts, grains seeds and feeds given in balanced small quantities twice daily, but supplemented with grass and legume mixtures and abundant water. Mineral supplement should be given regularly. Breeding and rearing of the young are essentially natural, and the baby rabbits should be allowed free access to their mothers until weaned. Disease control is essentially the same as with other animals.

## 6.3 NON-HERBIVORES

### 6.31 PIGS

Pigs are very important in West Africa (a) because there is a strong belief among the Muslims that pig eating is forbidden on religious grounds, and (b) because food processing in West Africa advances with pig products (e.g. bacon, ham, sausages) much more easily. Nevertheless, pig farming on large commercial lines is still confined to the universities, research institutes and government farms. Only a few private pig-keeping ventures have succeeded, and the majority of the pig population in West Africa is confined to the south. This is because the exotic breeds of pigs in use in West Africa require huge amounts of food, good houses and paddocks, which most peasants cannot afford.

West African pigs are small in size, have a very small body, tend to produce a lot of fat if fed too much and are uneconomic. For these reasons exotic breeds (e.g. Large White, Large Black, Wessex, Saddleback, Tamworth) have been tried for cross-breeding with local breeds. Nowadays, these exotic breeds in their pure forms are used for breeding and for the commercial production of pork (meat), ham, bacon and sausages.

#### *Care of Pigs*

Pigs are omnivorous, feeding on virtually anything. It is estimated that nearly 60 per cent of the cost of raising pigs is on feeds alone. Pigs survive mainly on concentrates with about 20–25 per cent of protein. With the traditional methods pigs are fed as poorly as sheep and goats.

But unlike them, pigs cannot survive on herbage. Instead they roam about, feeding on kitchen wastes and anything else they can find.

But the exotic breeds have to be adequately fed. The average ration will include maize, guinea corn meal, palm kernel or cotton seed meal, groundnut cake, blood or fish meal, salt, and a bit of palm oil. Feeds are given twice daily except in the case of nursing mothers which are fed thrice daily. Because feeding pigs can involve huge amounts and the costs are so high, care is necessary to avoid wasting food. Underfeeding with pigs is as bad as overfeeding, since pigs tend to put on fat quite readily – a tendency which has to be discouraged if pigs are kept for bacon and ham.

Water is an important item in pig raising. It is also a source of several internal parasites and other infections if the water supply is dirty or contaminated. Pigs drink a huge amount of water to aid digestion. In addition, body temperature control is achieved by wallowing in cool water. Hence the supply of water for drinking and wallowing should always be regular, clean and cool. Parasitic worms (e.g.tapeworm) are their most common problem; hence pigs should be kept indoors as much as possible. Paddocks for exercise should be used on a rotational basis to reduce infection.

Pigs have a very strong snout (mouth, jaw and nose ridge – see Figure 5.5); this is often used in burrowing for roots and in cooler parts of the earth for grubs. Therefore houses and paddocks designed for pigs have to be strong to withstand their burrowing action. Pen floors have to be especially reinforced and concreted. The floor should of course be rough to avoid slipping and accidents. Pig pens should always have sleeping floors and a separate feeding area with feeding and water troughs. In the case of farrowing pens, farrowing rails should be provided. It is important to avoid draught and too cold floors as a check against pneumonia.

Avoid overcrowding pigs as this leads to a lot of unhealthy biting habits, which may lead to wounds, accidents and other unproductive consequences.

*Pig breeding*

The average adult boar (male pig) can handle over twenty sows in a year, and a good breeding boar is the key to successful pig raising. The boar is sexually mature at 7 months. His sexual performance is, however, influenced by the rate of his sexual development, health, mating behaviour, and the quantitative and qualitative aspects of his senses.

Good sows usually farrow (have piglets) twice a year and should on the average be able to wean eight to ten piglets per litter. Piglets for the first 4 weeks of their lives feed mainly on their mother's milk. The sow's teats must be kept clean, and there should be as many teats as there are piglets. Good breeding sows must have at least ten teats, which should be filled with milk at farrowing. Where piglets are more than the number of teats, or the milk supply of the sow is inadequate, foster mothers or artificial milk feeding should be done.

Piglets are usually weaned at 8 weeks and are mature in 6 months for the slaughter. They may be used for breeding at 9 months. Male piglets that are not fit for this purpose are usually castrated at 6 weeks and fatten into hogs at 5–6 months.

## 6.4 POULTRY

### 6.41 FOWLS

The most common of farm animals, and the most easily adaptable, are poultry, including fowls, turkeys, ducks and geese. In West Africa the indigenous poultry are very small, are poor layers of eggs and produce little meat, but they are hardy and resistant to local diseases and survive at virtually no cost of housing or food to the keeper.

It is therefore not surprising that in many parts of West Africa the keeping of exotic breeds of fowls is being practised extensively. Imported breeds (e.g. Rhode Island Red, White and Brown Leghorns, Austrolops, Plymouth Rock) and their crosses with local breeds are the most productive for commercial purposes. These breeds are kept either for eggs (e.g. Leghorns) or for poultry meat (e.g. larger breeds such as the Rhode Island Red and Austrolops). The dual purpose birds (e.g. Rhode Island Red, Plymouth) produce both eggs and a good quantity of meat.

Table 6.2 *Useful information on fowls in West Africa.*

| | *Native Fowl* | *Exotic* | |
|---|---|---|---|
| Breed | Mixed breed | Rhode Island Red | Leghorns |
| Production | Eggs & meat | Eggs & meat | Eggs |
| Egg colour | Dirty white | Brown | White & Brown[a] |
| Egg output (ann.) | 40–60 | 150–200 | 200–250 |
| Body weight in kg (lb): | | | |
| male | 1·0–1·5 (2·2–3·3) | 4·0 (8·8) | 2·7 (5·9) |
| female | 0·5–1·0 (1·1–2·2) | 2·7 (5·9) | 2·0 (4·4) |
| Egg incubation: | | | |
| natural | Yes (20–21 days) | No | No |
| artificial | Yes (20–21 days) | 19–21 days | 19–21 days |
| Management system | Free range | Deep litter | Battery |

[a] White Leghorns lay white eggs and brown Leghorns lay brown eggs.

Eggs from fowls are either white or brown shelled, and the meat is white or brown (yellow). Some useful particulars about fowl breeds in West Africa are given in Table 6.2.

*The Egg*

The most important poultry product is the egg (see Figure 5.12), the quality of which determines the type of chicken produced on the farm. The egg is also a complete food containing a high degree of protein (*albumin*) and the *yolk*, which is a rich food reserve primarily meant for the new chick. The chick is represented in the egg by the *germ spot,* which has of course to be fertilised.

The quality of the egg is determined largely by: (a) the breed; (b) food, with high protein content and carotene, vitamins and minerals; (c) age of bird – birds over 5 years old are not economical food layers; (d) system of management; and (e) health of the bird. For good quality eggs at all times, management efforts should be made to make sure that these factors are given due and constant consideration.

With respect to eggs, *incubation* is the process necessary for the development of the fertilised germ spot (or ovum) into a chick. (Incubation in respect of diseases, on the other hand, is the phase through which the disease-causing organism has to pass before the first appearance of the disease symptoms.) Incubation period varies with different birds, as shown below:

| | |
|---|---|
| Fowls | 19–21 days |
| Turkeys | 25–30 days |
| Ducks | 25–32 days |
| Geese | 28–34 days |

Two incubation methods, the natural and artificial, are common practices in West Africa. Most native fowls and their first generation of crosses with exotic breeds still retain the instinct for natural incubation.

*Natural incubation*

During natural incubation the hen sits on (*broods over*) the eggs, which may be hers or another hen's. During this period warmth is supplied to the eggs from the hen, and aeration is also achieved naturally as the hen moves out from time to time to feed and relieve herself. On returning, the eggs are tilted over by the mother, and the heat impact is evenly distributed to all parts of every egg. Where there are more eggs than she can cope with – most hens can conveniently cope with ten – the extras should be removed; otherwise they will not be adequately treated, and many chicks may either fail to develop fully within the time period or die in the shell.

Once the hen initiates brooding, the brooding nest should be prepared from dry straw and soft tissues or even soft, clean, dry sand. During the brooding the hen should be fed and supplied with good, clean, cool water regularly, as often as she comes out of the nest. Protection against pests (e.g. termites, mites, lice) and dangerous animals (e.g. cats, wild dogs) should also be provided.

Immediately the chicks are hatched, they and the mother hen should be removed to a different nest, pen or cage, where the chicks should be fed on small amounts of fine clean sand and protein-rich concentrates in tiny quantities. The grit aids digestion in the gizzard of the chicks. The hen and chicks should on no account be separated until all hatching is completed. This will become obvious, as the hen of her own accord will move out of the nest with the chicks to signify the end of the incubation. Immediately after this the old nest should be burnt or buried, along with any egg remnants not hatched out.

*Artificial Incubation*

Artificial incubation is the technically improved method whereby any number of eggs can be incubated, following as closely as possible to the natural method. The number of eggs that can be incubated depends of course on the size of the machine (*incubator*) used. To obtain the best results from artificial incubation, the following activities have to be carried out as well as possible.

(1) *Control of temperature and humidity:* The temperature required in the incubator for fowl eggs is 37·5°C (99·5°F), with a change of not more than 0·5°C (0·9°F) either way. Humidity must go along with temperature to provide a balanced and ideal condition in the environment surrounding the eggs. In the first 17 days humidity in the incubation is best kept at 58–60 per cent, and gradually increased in the 18th day to about 90 per cent for the last week of incubation. Smaller incubators will perform better at lower humidity.

(2) *Ventilation:* Because eggs are living organisms, they respirate; therefore they need oxygen and produce carbon dioxide. Air (oxygen) should therefore be provided. It helps to place incubators away from walls to allow for the free circulation of air (oxygen and carbon dioxide).

(3) *Turning the eggs:* This should be done about 5–8 times every day until the 18th day, when turning ceases until hatching is completed. Smaller incubators are turned by hand and should be turned less frequently (3–4 times a day is enough). Some incubators have automatic turners for turning the eggs without it being necessary to open the egg chamber.

(4) *Candling:* This is the process by which the growth progress of the embryo in the egg is determined by the use of light. It is done a week after incubation. Infertile and cracked eggs and dead embryo that are transparent to light should be removed.
(5) *Disease control and prevention:* Control methods, especially for salmonella and other infectious diseases, should be carried out before and after every incubation, by fumigating the incubator with formaldehyde or any other disinfectant fumes.

*Management of Chicks*

*Brooder houses* are provided for chicks hatched artificially. In the brooders, warmth is supplied by the use of lamps (electric, kerosene, etc.), and the temperature is kept at 30–32 °C (86–89·6 °F) with humidity at 54–56 per cent, in the first week. Heat supplied artificially should be gradually reduced after the first week, so that by 6 weeks old there is no further need for the heating lamp. After 6 weeks the chicks are old enough to go into a pen.

When they are 6 months old they are ready for breeding (pullets and cockerels) or for the table as capons (castrated cocks). The sex of chicks can be determined at 1 day old – a process known as *sexing* – by examining their cloaca or by the genetic inheritance of sex-linked coat colour if these are known.

*Management of Adult Birds*

Four management systems are known for raising poultry, the free range, fold, deep litter and battery systems.

The *free range system* is the most common in West Africa for peasant production involving the local breeds. No food or housing is supplied, and the birds are always on their own and free to roam.

The advantage of the free range system is its cheapness: lack of any cost except purchasing the breeding stock. Disadvantages are many, but essentially it is a wasteful system. With more and more population pressure on the land, the fowls will soon have no space to roam.

The *fold system* is where the birds are kept in movable houses usually attached to a run where the birds are fed and supplied with water and succulent (soft) greens. It is a semi-intensive system.

Advantages of this system include:

(1) Birds are protected from rain, cold, direct heat, wild animals and pests.
(2) Feeding is improved and controlled, and other treatments are easily carried out without disturbing the birds more than necessary.
(3) Manure and droppings are supplied to the land, and when the pen is moved the land can be used for the cultivation of crops.

(4) Eggs laid are easy to collect as they are not left scattered all over the place – a common occurrence in the free range system.
(5) Chickens hatched from natural incubation can run along in the pen with their mother.

However, there are disadvantages too:

(1) The cost of the house can be an expensive item for poor peasants, even though cheaper houses which are just as good are now possible.
(2) The rapid spread of diseases may be encountered, but such pens can be isolated as a unit and treated.
(3) Production is poor as food may not be enough; fighting may become a constant feature.
(4) Cleaning of pens and general care will also cost some money, for labour and other expenses such as disinfectant, etc.

The *deep litter system* is a more intensive system than the fold system. The birds are kept indoors all the time on a pile of litter of sawdust, wood or cut straw and are fed and treated in the house or pen.

The merits of this method are:

(1) The supply of micro-organisms in the litter and the breakdown of organic matter supply some supplementary vitamins and minerals to the birds.
(2) It is a cheap method with respect to land and labour.
(3) Control of egg laying and other activities are easily handled.

Its disadvantages include:

(1) High infection rate may result from bacteria if the litter is too wet or disease control is not adequate.
(2) House may get too warm or too draughty, leading to cold and poor growth and production.
(3) Lack of exercise on the part of the birds is sometimes bad.
(4) Additional greens have to be supplied, thus increasing cost.
(5) Droppings (putrefaction) may generate too much ammonia as well as encourage the spread of diseases.

The *battery system* is a very intensive system. Birds are kept in battery cages especially designed so that they spend all of their productive life in one place, feeding, excreting, fattening and laying eggs in the same cage. Draught and too intense heat and sunlight must be avoided while arranging the batteries.

The advantages of this system are:

(1) Little space is needed, and labour in handling birds is reduced to a minimum.
(2) It is cheap to maintain since labour for supplying food, collecting eggs and the occasional sweeping of excreta and waste is very small.
(3) Disease control is very easy.
(4) Relatively higher yields of meat and eggs are guaranteed.

The disadvantages include:

(1) Costs of material for the cages are high, and cages sometimes have to be imported.
(2) Expensive foods are used, since the food is the only source of supply of the essential nutrients, including vitamins, minerals and antibiotics.
(3) The caging of birds is of course unnatural and, coupled with lack of exercise, may be detrimental to the health and performance of the birds.
(4) The quality of the eggs laid is usually low, and the eggs are not fertile since no mating with the cock has been done; the method cannot therefore be used with birds for breeding purposes.

On the whole, the system to be used will depend on whether production is for meat or eggs or both. The battery system is most suitable for eggs and meat. Where breeding has to be done on a commercial basis, the fold or deep litter system is advisable. The free range system is not recommended for exotic breeds.

### 6.42 DUCKS

Ducks are birds which also produce eggs and good meat for human consumption. They are not as many or as popular in West Africa as fowls. The most common local ducks on peasant farms are very small in size and poor layers. Since they, like fowls, are also raised on the free range system, they are not economic for large-scale commercial production.

Imported breeds (e.g. Khaki Campbells) are found on many agricultural research stations, universities and government farms. These exotic ducks under good care and management can lay up to 250–300 eggs a year, even though duck eggs, which are bigger than fowl eggs, are not very popular. Most people do not like the meat, except occasionally, because it tends to be fatty or greasy. The management, feeding and breeding features of ducks are similar to those of fowls, and they can be treated the same.

### 6.43 TURKEYS

These are much bigger birds. A full-grown turkey is probably three times the weight of a mature fowl, i.e. 10–15 kg (22–33 lb). Turkeys also lay eggs bigger than those of ducks, but their eggs are subject to the same prejudice as duck eggs. Turkey meat, on the other hand, is very popular although relatively more expensive. Turkeys also have useful economic advantages as presents during festivities and for special dinners at Christian festive periods (e.g. Easter, Christmas) and at the New Year in West Africa.

Turkeys are difficult to raise, have low fecundity and poor fertility, and are more expensive to feed as they are bigger and feed mainly on concentrates. All the management, feeding, breeding and health problems discussed for fowls also relate to turkeys, which can be considered as similar in very many respects.

## 6.5 FISH

Fish is important because it contributes about 1 per cent of the total food supply of the world. West Africa is blessed with a large coastal belt and a good number of inland waterways for fishing. The development of rural fish ponds is also becoming important enough for fish to become part of any agricultural study, even if only briefly.

The place of fish in food consumption in West Africa is given in Table 6.3, which shows that fish contributes well over a quarter of the total animal protein consumed in West Africa and the Sahel and well over half that in the coastal South Zone.

The fishery resources are also important because over three-quarters of the surface area of the world is water – resource base for fishes. In West Africa, fish come from: (a) fresh water, (b) salt water (marine fish), and (c) fish ponds.

When processed, fish reaches consumers in West Africa in many different forms, as: (a) fresh fish, (b) smoked fish, (c) sun-dried fish (*bonga*), (d) salted fish, (e) stock fish (imported from Norway and Iceland), (f) frozen fish (from the trawlers along the coast), (g) tinned fish (e.g. tuna, salmon), and (h) fish meal (an important ingredient of poultry ration). Senegal is one West African country that has developed a successful fish meal industry for export.

Table 6.3 *Proportion of fish in food consumption in West Africa* (as % of food consumed 1960–2).

| | *Savanna Zone* | *South Zone* | *South west Zone* | *Nigeria* | *Cameroon* |
|---|---|---|---|---|---|
| Quantity of fish consumed: kg/yr (lb/yr) | 10·2 (22·4) | 12·7 (27·9) | 12·0 (26·4) | 5·8 (12·8) | 13·2 (2·0) |
| Calories from fish/day | 19 | 33 | 28 | 9 | 22 |
| Protein from fish gm/day (oz/day) | 3·4 (0·12) | 4·1 (0·14) | 4·2 (0·15) | 1·4 (0·05) | 3·2 (0·11) |
| Fat from fish gm/day (oz/day) | 0·3 (0·01) | 1·6 (0·06) | 1·2 (0·04) | 0·4 (0·01) | 1·0 (0·04) |
| Per cent of total food from fish with respect to: | | | | | |
| – calories | 0·9 | 1·6 | 1·3 | 0·4 | 1·0 |
| – fat | 0·8 | 6·5 | 3·4 | 1·0 | 2·1 |
| – total protein | 4·9 | 8·9 | 10·0 | 2·3 | 6·3 |
| – Animal protein | 27·2 | 56·9 | 67·7 | 21·5 | 40·5 |

*Source: Agricultural Commodities: Projection for 1975 and 1985* (FAO, Rome)

*Methods of Catching*

In most West African countries, catching fish is still largely traditional. Fish are caught by a number of methods, including: (a) by hand, (b) baskets and traps, (c) fishing spear, (d) fishing nets (from canoe), (e) poisoning, and (f) recently by use of trawlers and fishing boats.

Fish farming, by use of pounds, is becoming more and more important. Fish, like crops and other livestock discussed earlier, can be 'cultivated', especially in inland rivers, lakes, coastal swamps, lagoons, dams, reservoirs, paddy fields and village and family ponds. Improvement in fishing boats, fishing gear and fishing techniques have still to be done, because the fish industry was neglected for a number of years. These improvements when achieved will clearly improve the quantity and quality of the fish catch, thus improving the protein diets of many people in West Africa.

*Methods of Processing*

In very warm climates, such as obtain in West Africa, fish decomposes rapidly. At temperatures of over 30 °C (86 °F) some may become so spoilt as to be unacceptable within a few hours after catching (death) unless preservative methods are applied. This is why processing is so very important. Hence processing is carried out for three main reasons: (a) to preserve fish from deterioration and decay; (b) to improve the quality (aroma, taste, etc.); and (c) to facilitate improved packaging, storage, transportation and distribution.

The important processing methods of general application in West Africa include: (a) chilling, (b) freezing, (c) smoking, (d) salting, (e) drying, and (f) the production of fish products and canned fish (e.g. fish paste, fish pie).

*Chilling* is usually done at 0 °C (32 °F). In some cases the fish may be *filleted* (have the intestines, head, tail and scales removed and be cut into pieces) before chilling. Machine filleting is sometimes done, and the machine must be cleaned regularly to avoid the spread of bacteria and other micro-organisms that may contribute to decay and putrefaction of the fish later. Chilled fish can be transported over very long distances and will keep in good condition for months.

*Freezing* of fish is becoming more and more important in a number of big cities of West Africa. After freezing, the fish is kept in cold storage and distributed in that form until it is needed for food. Then thawing (i.e. getting the fish back to room temperature) has to be done before the fish is cooked or processed for consumption. To thaw the fish, it is left either in the air or over warm air or warm water. It is essential that, during thawing, care has to be taken to avoid secondary stage infection and/or contamination immediately after thawing. A good deal of the earlier opposition to frozen fish in West Africa was due to its rapid decomposition and deterioration after thawing.

One of the most common and easily feasible methods for preserving fish from canoe (peasant) fishermen (small scale) is *smoking*. In addition to preservation, smoking gives flavour and colour as well. For good smoked fish, the fish is first soaked in salt solution (*brine*) for only a few minutes. Sometimes the brine solution is simply sprinkled on the fish while it is being smoked. Smoking is best done at temperatures of 15–30°C (59–86°F) for cold smoking (Yoruba – *gbèdè*) or at 90°C (194°F) for hot smoking (well-dried). Smoked fish does not keep for very long. In fact, it needs to be constantly reheated, or sun-dried at least once a week, and kept well aerated but away from flies and other contamination if it is to remain fit for consumption.

*Salting* is useful where the fish is likely to be used for food (e.g. salads) without cooking. Some salted fish are subsequently sun-dried or even smoked. Salt concentration of 9–10 per cent sodium chloride ($NaCl$, table salt) is effective.

*Bonga* fish produced from inland water and lake fish are generally processed by *drying* (usually sun or hot air drying). Filleting to remove the gut contents helps to reduce infection during drying, which may last a number of days.

*Canning and manufacturing* are the most common commercial practices. Fish canning is a specialised field and can only be practised when many tonnes (tons) of fish of the same type (e.g. sardines, salmon, tuna, shrimps) are involved.

### *Feeding or Cultivation of Fish Ponds*

The best source of information on the establishment, inoculation, management and cropping (or harvesting) of fish ponds or on fish cultivation is to approach a local fishery officer in the Ministry of Agriculture and/or Natural Resources.

Basically, fish will eat virtually anything in water provided it is soft enough. But under conditions of cultivation, especially in artificial ponds and inland swamps, food has to be provided in the form of human and animal wastes. In order to avoid contamination, such wastes have to be treated to destroy pathogenic organisms. In addition, a number of protein-rich foods recommended for livestock (e.g. blood meal, bone meal, fish meal, groundnut cake, maize meal) are all good food for fish. Depending on the type of fish in the pond and their numbers and ages, the food should not be too much. Rather it should be in small quantities supplied at frequent and regular intervals – three or four times daily. It will, however, be necessary to go into more detailed readings on fish culture and to seek expert advice before embarking on being a fish farmer.

*Part Four*

# THE USE AND MAINTENANCE OF FARM EQUIPMENT, MACHINES AND STRUCTURES

*Chapter 7*

# Farm Tools, Equipment and Machines

## 7.1 CLASSIFICATION AND FEATURES

### 7.11 CLASSIFICATION

Farm tools, equipment and machines are classified according to (a) the kinds of operation they perform on the farm, and (b) the stage of cultivation at which they are employed.

Those implements used in the preparations for planting are called *primary (1°) implements* because of their primary role in cultivation. Examples of these include cutlasses, hoes, rakes, mowers, ploughs and harrows.

Those which come into use during the planting operations and maintenance of the crops during growth are referred to as *secondary (2°) implements*. These include planters, weeders, shears, secateurs, sprayers, fertiliser-applying devices, etc.

Farm implements that are used during the harvesting and stacking of crops are referred to as *tertiary (3°) implements*. These include mattocks, mechanical harvesters, harvesting machines (e.g. combine-harvester), shellers, peelers, etc.

A host of other implements fall into a fourth grouping of *quaternary (4°) implements*, and they are used for after-harvest purposes such as grading, weighing, grinding, etc. Examples of these are scales, graders, shellers, peelers, choppers, separators, driers and grinders (or comminutors).

### 7.12 FEATURES

Since farm processes are necessary and delicate, farm implements are designed with great care and accurate specifications to adapt them for the uses for which they are intended.

Most of the implements can be used for most crops. The primary implements can be used for almost all types of crop, the major limiting factor being the soil type to which they are to be applied. Since in West Africa there is a noticeably high range of soil types, most farm implements are designed for use on a variety of soil types. Consequently, each implement is made in different forms for use on different types of soil.

With most secondary implements, the soil type has little or no significance. Instead it is the type of crop planted (or to be planted) that dictates the form of implement used. For implements such as planters, the seedling rate of the crop to be planted, coupled with the growing habit, dictates the type of planter to use, the spacing and the setting to be adopted on the implement. Fertilising devices, sprayers, weeders and most other secondary implements are also affected by the growing habit of the crop, and are as such designed to be able to cope with different kinds of crops. For example, there is the simple sprayer, which is useful when performing localised spraying on short and well-spaced crops. There is also the general pressure sprayer, which has a very much longer stem and utilises pressure for its functions. It can be used for short or tall crops, for well-spaced or closely packed crops, for localised or total spraying, and for a relatively very much longer period of time (see Figure 7.2).

Tertiary implements are mostly affected by the type of crop and the growing habit of the crop. In their case the soil type is of no importance. These implements are as such designed in different forms for use on different types of crops. For example, the mattock – used for harvesting cocoa – has a long handle with very sharp and rigid cutting blades at the head for the high and firmly attached cocoa fruits (see Figure 7.1n); while the sickle – used for harvesting closely packed, long stem crops (e.g. hay, millet) – has a single curved blade with a very sharp edge, and a short handle (see Figure 7.1m).

## 7.2 FARM TOOLS AND EQUIPMENT

### 7.21 DEFINITIONS

Farm *tools* are implements of very simple nature which are commonly found with farmers who cultivate small farms. They are made up of a few simple parts (usually metal and wood) which are attached to one another. They have little mechanical advantage because the amount of work they do is dictated by the effort of the farmer. Examples of farm tools are: cutlasses (machete or field cutlass), hoes, rakes, forks, diggers, spades, shovels, trowels, shears, knives, sickles, secateurs, mattocks, etc. (see Figure 7.1 a–n).

A farm *equipment*, on the other hand, is an implement made up of two or more parts which work together for a common purpose. The farm equipment can be a simple mechanical device (e.g. the horse-drawn plough) or a more complex device (e.g. the sheller). A farm equipment differs from a farm tool in that it requires less effort from the farmer for operation than does the farm tool. Some examples of farm equipment are: ploughs, harrows, tillers (which can be tractor-, mule- or horse-drawn), some sprayers, shellers, scales, irrigating devices, etc. (see Figure 7.2 a–b).

## 7.22 USES

### *Tools*

*Cutlasses* are used for: (a) cultural practices such as weeding, planting, thinning, roguing and harvesting; (b) post-harvest practices such as the breaking and cracking of pods and shells; (c) domestic purposes such

Figure 7.1 *Some farm tools in use in West Africa.*

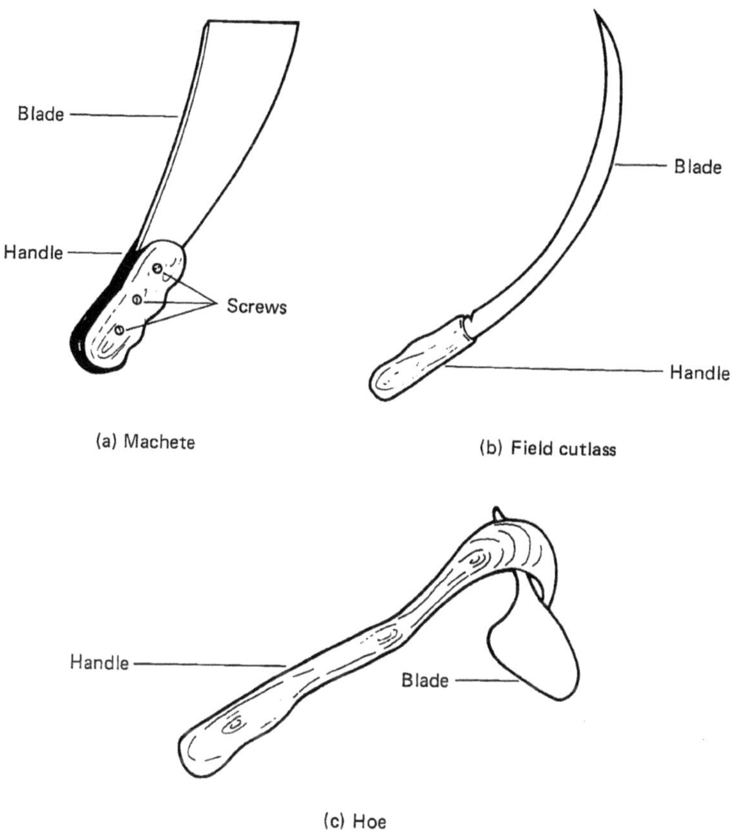

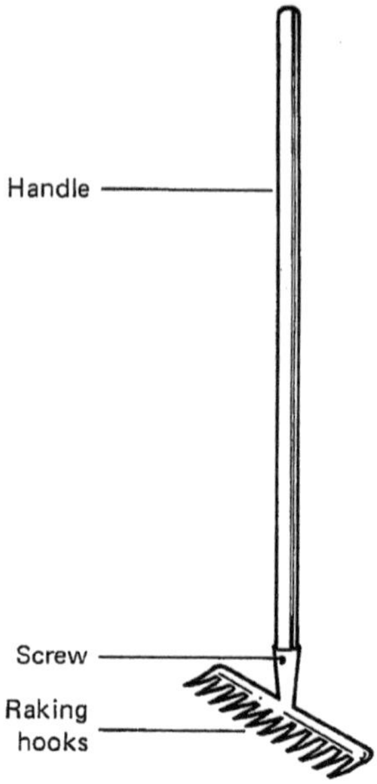


(d) Rake

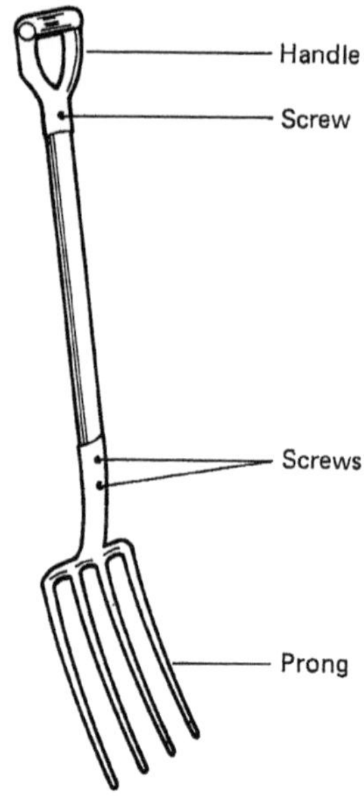


(e) Fork

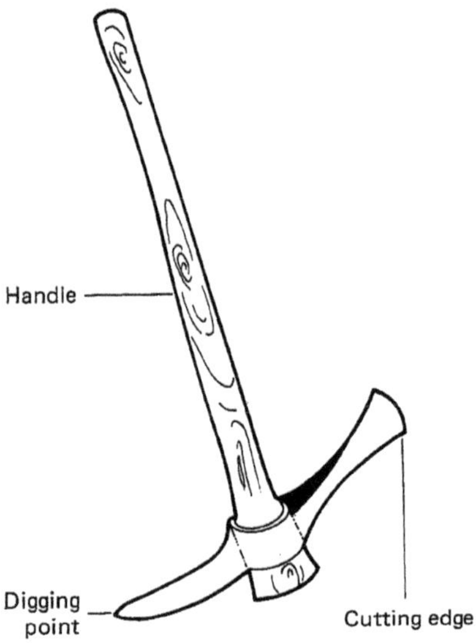


(f) Digger

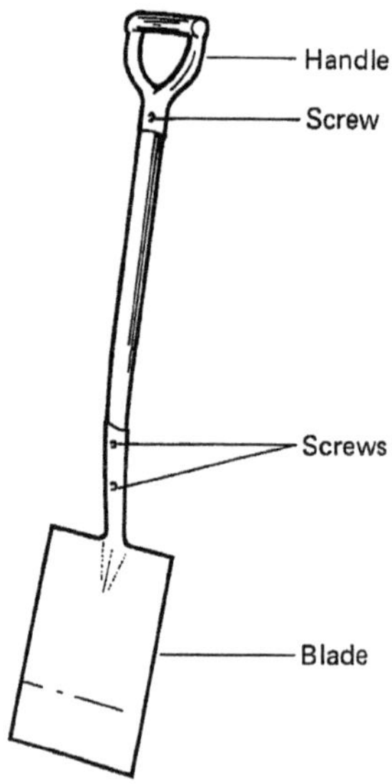


(g) Spade

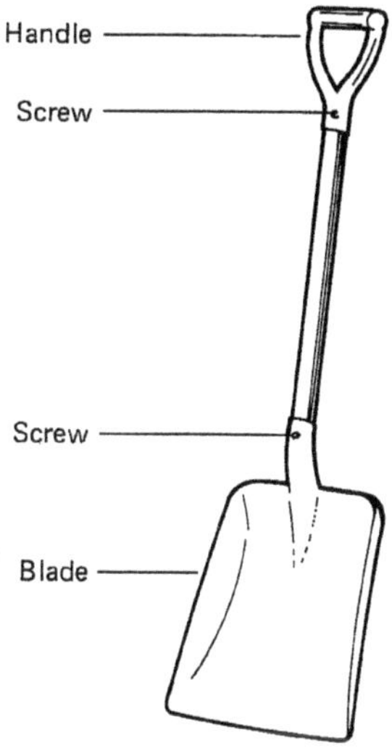


(h) Shovel

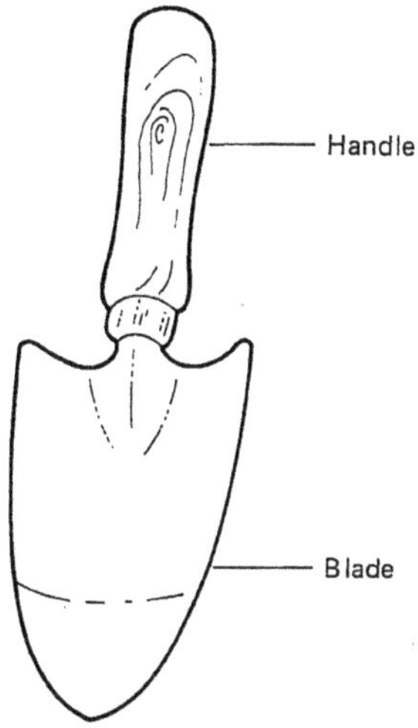


(i) Trowel

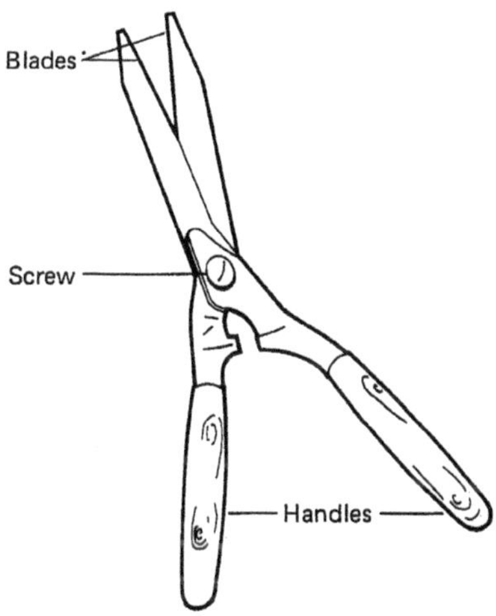


(j) Shears

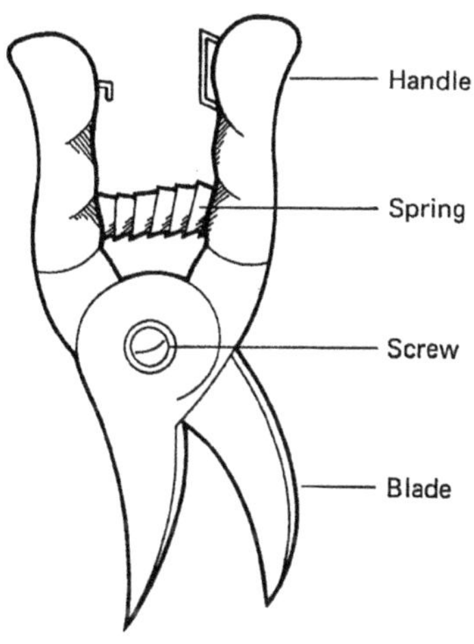


(k) Secateurs

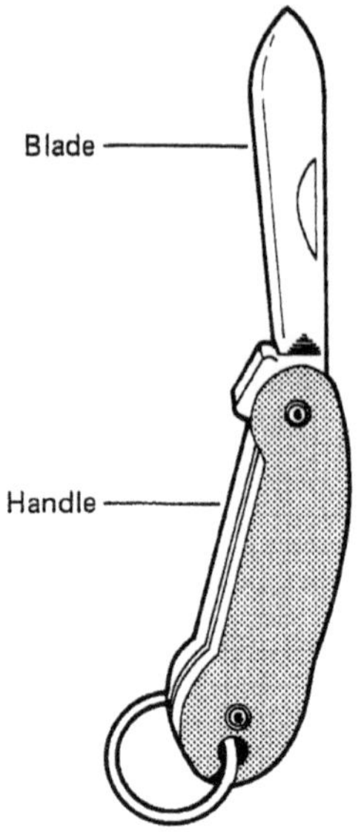


(l) Budding Knife

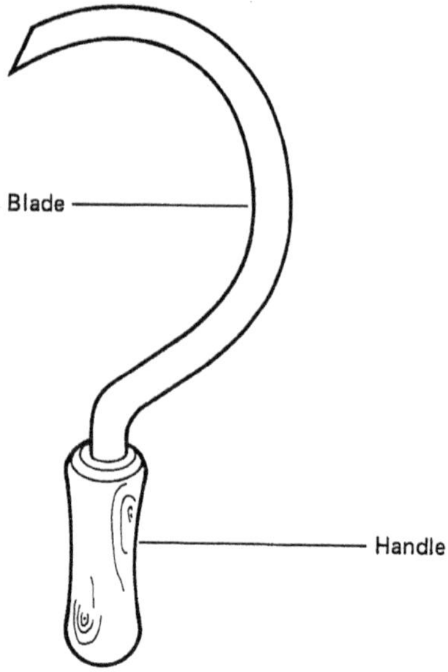


(m) Sickle

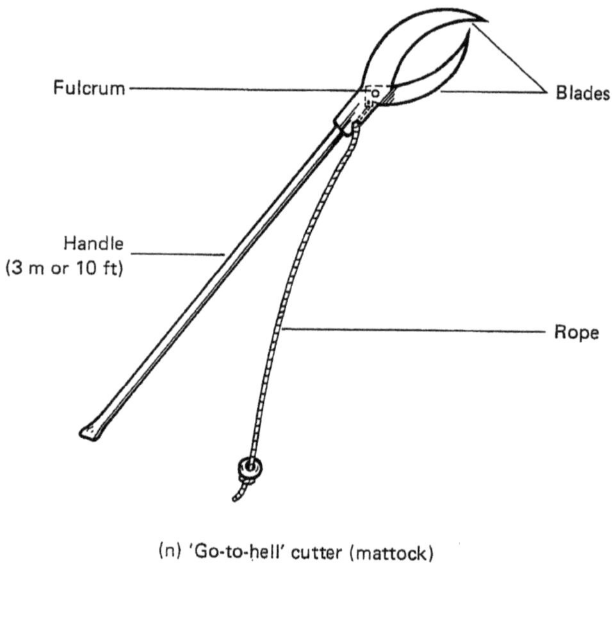


(n) 'Go-to-hell' cutter (mattock)

as the cutting of firewood and of pieces of meat; and (d) digging small pits and holes.

*Hoes* are used for: (a) making beds and ridges; (b) cultural practices such as weeding and planting; (c) digging holes and making drains and irrigation canals; and (d) transplanting.

*Rakes* are used mainly for: (a) cultural practices such as the preparation of a clean field by digging into soil to gather up all weeds and other foreign materials present in the soil; (b) weeding; and (c) collecting seeds and other loose farm refuse.

*Forks* are very useful for: (a) turning the soil to loosen any hard surface; (b) removing some 'die-hard' weeds (e.g. perennial weeds); and (c) annual and compost turning.

*Diggers* are used for: (a) digging deep into soil during the construction of structures such as drains, gutters and building foundations; and (b) loosening a hard topsoil.

*Spades* are used for: (a) digging and moving loose soil; (b) making beds and ridges; (c) weeding; (d) transplanting; (e) mixing feedstuffs for livestock; and (f) cutting roots.

*Shovels* have relatively weaker blades than spades. They are used for: (a) digging and moving loose soil; (b) moving feeds, grains and fertilisers; and (c) transplanting.

*Trowels* are used mainly in the garden for: (a) digging and moving small quantities of soil; (b) moving grains, feeds and fertiliser materials; and (c) moving seedlings during transplanting.

*Shears* are used for cutting leaves and branches during thinning.

*Secateurs* have sharp, scissor-like blades. They are used for: (a) pruning and trimming leaves and branches; (b) harvesting some crops, such as vegetables (e.g. spinach).

*Knives* are used for: (a) budding; (b) cutting during harvesting and post-harvesting practices; and (c) domestic purposes.

*Sickles* are commonly used for harvesting hay, dried bedding grasses, and other tall grasses (e.g. millet, sorghum) for silage making.

*Mattocks* are used for harvesting crops on tall plants (e.g. cocoa, kola, citrus, guavas, avocado and other fruit trees).

*Equipment*

*Tillers* are primary tillage implements used for turning the topsoil. They can be tractor-, horse- or mule-drawn.

*Ridgers* are primary cultural implements used for making ridges. They come in different kinds and sizes and can be tractor-, horse-, or mule-drawn.

*Harrows* are used: (a) for breaking up large soil particles, to facilitate further cultivation; and (b) for final preparations of the seed beds.

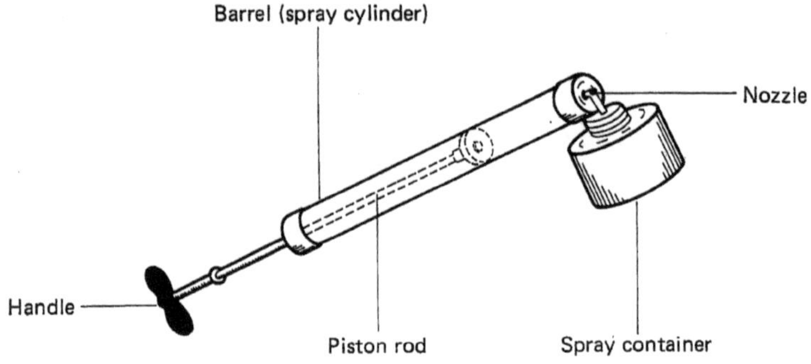


(a) Simple sprayer (for very localised spraying)

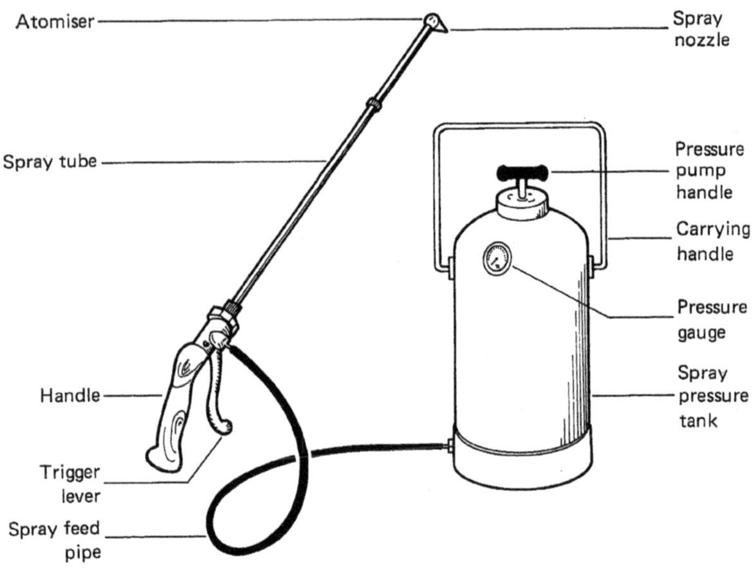


(b) General purpose sprayer

Figure 7.2 *Sprayers used in West Africa.*

*Mowers* are used: (a) for weeding; and (b) on fields for controlling the growth of grasses to certain levels.

*Harvesters* are used to harvest crops. They come in different kinds, and the most popular of them is the combine-harvester which combines the harvesting, threshing and winnowing processes.

*Sprayers* come in different kinds and sizes. Some are used as water sprayers in vegetable and horticultural gardens, while others are used to spray pesticides and fungicides.

### 7.23 MAINTENANCE

The usefulness and length of the useful life of any farm tool or equipment depend on the quality and proper handling of the implement. Lack of adequate maintenance and of good handling during use lead to a great reduction in the usefulness of any implement. This may lead to constant replacement, which gives rise to higher costs of production of farm goods. The following rules pertain to all farm implements:

(1) Handle with care at all times.
(2) Use in the way directed by the manufacturer. Most manufacturers enclose instruction manuals or handbooks with machines and equipment. This must be read thoroughly and the instructions followed carefully if the machine is to be used effectively and for as long as possible.
(3) Thoroughly clean and store well after use, preferably in locked or guarded buildings.
(4) Check regularly if not in use, and treat (e.g. lubricate with oil to protect it from rust, or dust) according to the directions of the manufacturer.
(5) Keep away from weather and other hazards (e.g. rain, fire, strong chemicals).

## 7.3 FARM MACHINES

### 7.31 SOURCES OF FARM POWER IN WEST AFRICA

When we speak of power on the farm, or farm power to be more exact, we refer to the capacity for exerting or applying mechanical force. Several farm operations (e.g. planting, weeding, transplanting, harvesting, building and repairing roofs and other constructions) can only be done with the application of mechanical force. Therefore the farmer needs farm power, and in several forms.

Some of the more common sources of farm power in West Africa are as follows:

(1) *Human power* is provided by the farmer, his family or hired labour.
(2) *Animal power* is used for ploughing, weeding, carrying and transporting.
(3) *Mechanical power* is obtained through the use of machines (e.g. tractors, harvesters).
(4) *Electrical power* is used for moving other machines and for light and heat.
(5) *Water power* is not commonly used in West Africa, but is helpful in transportation, timber logging and the provision of electricity and irrigation water through the construction of dams.
(6) *Wind power* is not very important in West Africa either, even though most of the post-harvest farm operations (e.g. the winnowing and sifting of grains, cowpeas, etc. and such related activities) are based on wind power; more complicated forms of wind power such as the windmill are more common.
(7) *Sun power* is not directly in use as such, but indirectly the sun is used for drying operations at virtually no cost to the farmer, except for his labour in exposing farm products for sun-drying.
(8) *Wood/charcoal power*: As a form of energy and heat generation, firewood and charcoal are important and cheap although not used directly as a source of farm power, even though they could be so used.

## 7.32 USES AND CARE OF FARM MACHINES

Most farm machines require little or no manual operation because they are motorised (motor-engined). They come in different kinds and forms for different uses. The common ones on the farm include: (a) tractors, (b) sowing machines, (c) fertiliser spreaders, (d) harvesters (e.g. combine-harvester), and (e) processors (e.g. shellers, peelers, mills, extractors, choppers).

### *Uses*

*Tractors* come in many types, and the model chosen depends on the purpose for which it is to be used. Tractors are used for pulling implements and other machines during farm operations. When fitted with a trailer, they are also used for transporting farm inputs, products and farm refuse.

*Sowing machines* are used for planting seeds. They can control the depth, the spacing and the quantity of seeds sown.

*Fertiliser spreaders* are tractor-mounted machines used for the sole purpose of spreading fertiliser material on the field. They come in different models and types.

*Harvesters* are machines used for harvesting some crops. The most popular of the harvesters is the combine-harvester, which harvests, threshes and winnows the crop, mostly cereals.

*Processors*, as the name implies, process the harvested crops:

(1) *Shellers,* for example, are used for removing maize from the cob.
(2) *Peelers* remove the outer cover of root and tuber crops (e.g. yam, cassava, potatoes).
(3) *Extractors* are used for pressing out the fluid from crops; for example, the oil extractor presses out the oil from palm kernels.
(4) *Choppers* are used for cutting up crops into little bits, as done in the preparation of silage for farm animals. The chopper is run over the crop (e.g. maize, sorghum, millet), and the lower blade cuts off the crop, which is passed through a chute into a chamber where chopping is done. After chopping, the chopped material is passed into a container by pneumatic pressure, via another chute. The silage thus made is stored in the silo until needed for the feeding of animals.
(5) *Mills* are used for grinding seeds into flour. They are also called *comminutors* for they operate by comminuting the seeds. They come in different models and types, and the degree or comminution can be selected on the machine.

*Care*

The necessary care of machines involves both regular maintenance procedures and the care to be taken when using the machine.

Regular or routine care and maintenance involves the following:

(1) Keep all machines free from dust and dirt.
(2) Protect them from rust by keeping them away from rain and direct heat of the sun.
(3) Check all bolts and nuts regularly to see that they are properly oiled and tightened; loose nuts may give rise to bad performance and serious accidents, apart from damaging the machine.
(4) Grease and lubricate all moving parts, using the correct lubricant and according to directions.
(5) Keep all machines locked up or under guard when not in use to prevent their being stolen or damaged.

When the machine is in use, special care should be taken to:

(1) Check for oil, water, fuel, tyre pressure, etc. both before use, when

driving and after use, to reduce wear, tear and damaging of the essential parts.

(2) Handle gears, steering and other adjustable parts properly, to make productive use of the machine and to avoid damaging the delicate parts.

(3) Service regularly, to keep the machine in good condition for top performance and longer life.

*Chapter 8*

# Farm Buildings and Structures

## 8.1 FARM BUILDINGS

### 8.11 USES

Farm buildings are used for several purposes:

(1) They protect animals from poor weather, wild animals and thieves.
(2) They provide ease of access to the animals for care and attention.
(3) They ensure hygienic production as in a dairy.
(4) They improve yields of products such as eggs and fruits.
(5) They store and safeguard expensive equipment, materials and farm products.
(6) They may have special uses (e.g. offices, engine rooms, stores, dams).

### 8.12 TYPES

Farm buildings can be grouped under two broad subheadings; (a) temporary (or makeshift) buildings, and (b) permanent buildings.

*Temporary buildings*
These are buildings, sometimes prefabricated, made from timber, dried straw or reeds, tree bark and aluminium sheets. Sometimes mud, palm fronds and sticks are also used. The materials used for the temporary building depend on the climate and the purpose for which the building is constructed. The makeshift buildings commonly seen on farms include: tool and equipment sheds, storage sheds, construction sheds or workshops, garages for machines, nurseries and animal pens (e.g. cattle pens, piggery, poultry houses, abattoirs and dog kennels). Temporary buildings are simple constructions made from cheap, easily obtained materials, and it is thus easy to leave them when the farming site is changed. Most farmers prefer temporary buildings because they are cheap.

*Permanent buildings*
Permanent buildings are made from more costly materials and are less simple to construct than makeshift buildings. The materials used for

the construction of permanent buildings are mainly: cement, gravel, sand, stones, iron rods, wood, with asbestos and aluminium sheets and planks or wood for roofing. Permanent buildings require much deeper foundations and more careful construction and take more time to build than prefabricated buildings. Owing to the higher capital outlay in the construction of permanent buildings, it is not the usual thing to abandon the buildings when changing farm sites, and so they are found mostly on permanent farm sites.

Permanent buildings are erected when particularly needed for such purposes as storing fertiliser, feed, seed, farm products and pesticide, fungicide and other chemicals. Permanent buildings are also used for storing drugs and expensive tools and equipment. The housing quarters and offices on a farm are also permanent or semi-permanent buildings, while some expensive animals (e.g. bulls, racing horses, pedigree animals) have to be permanently housed, partly as a protection against thieves, bad weather and wild animals, partly to keep them away from indiscriminate mating or wasted energy and more essentially for care (feeding, grooming) and attention.

## 8.13 MAINTENANCE

### *Temporary Buildings*

The costs of constructing permanent farm buildings constitute a considerable proportion of capital outlay, and adequate maintenance is essential to keep these costs at the minimum level in order to increase farm profits.

Temporary buildings, on the other hand, require more regular care and maintenance than permanent buildings, for several reasons:

(1) Since they are constructed from wood, straw and other easily ignited materials, they must be protected from fire at all times, especially during the dry season when fire hazards are frequent.
(2) Termites and other pests and animals have caused the failure of many makeshift buildings by eating away the wooden parts. Consequently the woods used in the construction of such buildings must be protected against such attacks by painting them with chemicals (e.g. solignum), and by using seasoned woods.
(3) Rotten and rotting parts of the building must be changed immediately, and regular checks must be maintained to locate all the weakening parts of the building.
(4) Roofs leak more frequently, and strong winds can carry the roofs of makeshift buildings away. Hence, it is mandatory to secure the roofs firmly on the buildings and to change them as often as necessary. Moreover, the poles of the buildings must be well dug

into the ground and made very rigid. Preferably a strong foundation should be made for the building.

*Permanent buildings:*

(1) Prevent the incidence of fire by taking necessary care when handling naked flame and by keeping dry grasses, straws, etc. away from the building, especially during the dry season.
(2) Use seasoned woods for all the wooden parts and coat with solignum.
(3) Use the correct cement plus sand (or gravel) mix during the construction of the building, to ensure long life.
(4) Paint all iron members to protect them against rust, and block all leaking roofs immediately on discovery.
(5) Mend cracked walls, and change all rotten woods to prevent total collapse of the building.
(6) Install lightning conductors to prevent mishaps during electrical storms.
(7) When not in use, conduct regular checks to ensure that the building is in good condition.
(8) White-washing dairy pens keeps buildings hygienic and presentable. The use of disinfectants, insecticides and fungicides is often advised.

## 8.2 FARM STRUCTURES

Farm structures are constructions made on the farm site to enhance the smooth running of farm operations.

The types of structures found on a farm are dictated by the kind of farming practised and the conditions prevailing on the site or in the locality of the farm.

Structures commonly found on farm sites include: (a) drainage canals; (b) dams, water tanks and irrigation canals; (c) crushes and animal dips; (d) manure pits; (e) disposal pits (e.g. soak-aways); (f) electrification poles, wires and connections; (g) fences; (h) silos; and (i) abattoirs.

### 8.21 TYPES AND USES

*Drainage Canals*

A drainage canal is a gutter-like or dug-out structure for conducting excess water away from the farm to a central pool (probably a dam, a flowing river or an unused or uncultivated area). The canal may be cemented to prevent its increasing in size due to erosion. Drainage canals are commonly in waterlogged areas (e.g. valleys).

*Dams, Water Tanks and Irrigation Canals*

Dams, water tanks and irrigation canals are very essential in areas with low rainfall, and in other areas during the dry season when the amount of rainfall is not sufficient to provide the water requirements of the crops. *Dams* are structures made across a river or stream so that the flow of water is prevented or greatly reduced. As such, water accumulates in the dam, and the water is used for irrigating the crops.

The *irrigation canals* conduct the water to the crops. Depending on the location of the dam and the dimensions of the farm, a *water pump* may be incorporated with the dam. The pump sucks the water from the dam through large tubes and forces the water into the irrigation canals, which are constructed in such a way that the flow of water to the crops is smooth, by sloping the canals towards the crops.

*Water tanks* are used when the irrigation facilities cannot be employed (e.g. in the piggery, poultry house, cattle pens and for domestic purposes). Water may be stored in the tanks, which are placed high from the ground and can be filled with water from a tap or the rain or both. Tanks are made from aluminium materials to prevent rusting. The flow of water from such tanks is by gravitation and atmospheric pressure. The water can then be conducted to where needed by pipes.

*Crushes and Animal Dips*

Crushes and dips usually go together to reduce costs, as two structures in one. They are usually two areas of wooded paddocks between which is a narrow passageway (the *crush*). On one side of the narrow passage is a weighbridge for getting the correct live weight of the animals, either regularly or at the time of purchase or disposal. Taking the live weights of animals improves records of animals as well as their performance.

The other side or half of the narrow passage leads to a *dip*, which is usually a concrete bath sunk into the ground. The dip when in use is filled with water to which some disinfectants or insecticides have been added. When animals (e.g. cattle, sheep, goats, horses, donkeys) pass through, as of routine, the insecticides soak (or penetrate) through their coat to destroy ectoparasites (e.g. ticks, lice, fleas) on the animal. No big livestock farm can do without crushes and dips.

*Manure Pits*

Manure pits are holes dug out in the soil, into which waste plant and animal materials are poured in batches for the purpose of decomposing them to some degree before they are used on the farm as manure for crops. Such manure is incorporated into the soil prior to the planting of crops. The size of the manure pits depends on the size of the farm and the requirements for manure.

### *Disposal Pits*

Disposal pits (e.g. the soak-away) are pits into which useless farm wastes and domestic rubbish are dumped. They are often located at a reasonable distance from the housing quarters and offices, from where the stench cannot penetrate. Disposal pits are dug very deep into the soil and covered, so that they take a very long time before getting full. Human and animal faeces, urine and other useless wastes require to be permanently disposed of for cleanliness and health reasons.

### *Electrification Poles, Wires and Connections*

Electrification poles, wires and connections are often seen on big farms where most of the equipment and machines are operated by electricity, and where some delicate operations need to be performed after daylight. If the farm is located in a rural, non-electrified area, a generator may be installed to supply the voltage requirements of the farm. The poles and wires conduct the electricity to where it is required. Transformers, meters and switches are thus necessary for the completion of the circuits.

### *Fences*

Fences are specially designed barriers built around specified areas of land. Fences are made on the farm for several reasons: (a) to prevent animals from wandering out of the farm site; (b) to prevent dangerous animals and unwanted persons from gaining access to the farm; (c) to demarcate the boundaries of the farm site; (d) to control grazing on pastures; (e) to avoid damage to crops; (f) to ease the handling of animals for several operations; (g) to assist in training animals for shows (e.g. horses) or in certain habits.

There are several common types of fences:

(1) *Live fences* are very common in West Africa. They are cheap, made of vegetatively growing materials and easy to maintain, but they tend to overgrow if not properly maintained.

(2) *Barbed-wire-and-pole fences* are relatively more expensive to establish, but they are semi-permanent and do not require constant attention. The barbs (or metal spikes) can prove very harmful to animals by wounding them and may in fact damage the quality of skin or hide. Poles are of wood, iron or concrete.

(3) *Electric fences* can sometimes be used because of their effectiveness and ease of assembling and dismantling over varying areas of land. They are, however, very expensive to maintain, and the circuit may be easily fused or harm animals and young persons.

(4) *Stone, brick or concrete fences* are really permanent features. They are very expensive to build but last longer and are more effective if properly used.

(5) *Other fences* can be made of wood, tree bark, grasses and other materials. In general the quality, strength and effectiveness of any fence is dependent on the quality of materials used.

*Silos*

Silos are big, circular or rectangular structures made of stainless steel or concrete material and are used for storing grains and silage. The silo has a very big capacity, usually 50–100 tonnes (51–102 tons) and sometimes much more. It usually has just a port or two through which materials are put in and taken out. Huge quantities of grains and silage can be stored in a silo for a reasonably long period of time, due to the capacity and controlled conditions of the silo.

*Abattoirs*

Abattoirs are special housing structures constructed for the killing and butchering of animals (e.g. cattle, pigs, sheep, goats). It consists of a special passageway through which only one cow can pass at a time, and the passage leads to a slab on which the animal is killed after being securely held. Abattoirs are commonly seen on big cattle ranches.

## 8.22 MAINTENANCE

Since farm structures are essential for the everyday processes on the farm, lack of maintenance and the eventual breakdown of the structures may result in a general breakdown of farm operations and hence to a great reduction in farm profits. The upkeep of farm structures costs very little, and it entails just a routine check of each structure and the immediate repair of any damaged or spoilt part, whether the structure is in use or not.

Drainage canals should be checked to see that the canal is not blocked by weeds, waste materials, etc. Regular cleaning will take care of the canal. Dams, water tanks, water pumps and irrigation canals have to be checked regularly. The water pump must be serviced regularly, oiled, cleaned and fuelled. The tubes must not be allowed to become blocked by weeds and other materials. Irrigation canals must be cleared of obstructions. The dam must be checked regularly to see if it needs reinforcement, since it is under constant heavy pressure from the water mass accumulated. Water tanks must be cleaned regularly to prevent the sedimentation of dirt, which may lead to the eventual blocking of the conductor pipes.

Disposal pits must not be allowed to fill up. Regular addition of chemicals (e.g. caustic potash) or excavation helps to keep the waste material in the pit at a certain level. If these cannot help, other pits should be dug in good time to enhance a smooth switch-over to the new pits when the old ones become filled up.

Electric poles must not be allowed to rot. Such poles must be of high quality, seasoned and chemically treated. Any rotting pole must be changed immediately before it collapses, which may result in injury to people or animals and to a total disruption of the farm processes.

Electric wires must be long-lasting and must never be exposed from their plastic or rubber coverings. Since they conduct current whose voltage may be very harmful to man and animals, the wires must be replaced immediately if faulty or naked. The transformers and meters must be kept away from sight, preferably in small buildings specially built for that purpose. If a generator is used, the building must have adequate ventilation. Thus it should be a prefabricated building with a lockable door. Ventilation is necessary for the exit of the exhaust gases and smoke, which if prevented from escaping would choke up the generator and eventually contribute to the blowing up of the machine.

Fences must be checked regularly to see that no part has broken down. The wires or poles of the fences must be replaced immediately when found to be weak, rotten or snapped. Where live fences are used, the plants should be regularly trimmed and kept alive.

Silos must be kept very clean and airtight. Checks must be conducted for leaks regularly, to aid in the quick blocking of any existing leak.

The abattoir must be kept clean and the rigidity of the fences ascertained every now and then. The mechanical members must be oiled and cleaned regularly to prevent rust and malfunction.

*Part Five*

# MAKING PROFITS FROM FARMING IN WEST AFRICA

*Chapter 9*

# Keeping Simple Farm Records and Accounts

## 9.1 IMPORTANCE OF RECORDS AND ACCOUNTS

There are four main reasons why people take to farming. First and foremost, for food, shelter and clothing, i.e. for subsistence. Secondly, people cultivate the soil and keep livestock to have something for sale, gift or ceremony. Besides, gardening can be a hobby, and it may be interesting to keep animals for use as pets or for other reasons (e.g. hunting, house watching). Finally, it must not be forgotten that to many people farming is more than just a job: it is also a way of life in that some people would rather farm than do anything else; they simply like living on a farm.

For whatever reasons anyone undertakes farming, it is necessary to have a record which can serve to give the farmer all useful and relevant information about his crops, livestock and other possessions. The keeping of records and accounts on a farm is important for a number of reasons, some of which are: (a) to have an estimate of what the farm is worth at any time; (b) to give information for future planning; (c) to inform the farmer about the profitability or otherwise of the different units or enterprises on his farm; and (d) to inform the farmer about his resources use and output (product) relationship with a view to be able to appreciate his own work as a farmer.

## 9.2 TYPES OF RECORDS AND ACCOUNTS

The next questions are how many, in what form and how are records and accounts to be kept? The answers will depend on three important factors:

(1) *How big is the farm?* The bigger the farm area and the more the transactions undertaken, the more the number and complexity of records and accounts to be kept.

(2) *How many different units or enterprises are on the farm?* Ideally, it is far better to have a separate set of records and account book for every separate unit on the farm. The less profitable unit can be easily replaced with the more profitable unit.

(3) *What is the degree of commercialisation?* The more commercial-minded the farming business, the more detailed and accurate the records and accounts should be.

Whatever the level and complexity, the principle of keeping records and accounts is the same. Records have to be complete (i.e. include everything necessary), kept regularly (daily or weekly), accurate, and constantly checked and cross-checked. On the whole, any farm will require a minimum of six major sets of record and account books: (a) Labour Use Book; (b) Farm Purchases and Sales Book; (c) Crop Yield and Disposal Book; (d) Livestock Record Book; (e) Livestock Products and Disposals Book; and (f) Cash/Bank Account Book. These books could in some cases be called by different names, depending again on the emphasis placed on the different farm operations. For instance, a dairy farm might need a special Milk Production and Sales Book as well as a Breeding Record Book, or a kola–oil palm–cocoa farm might prefer to open a separate book for each of these major crops.

*Labour Use Book*

An example of the type of useful information this type of record book must have is given in Figure 9.1. The columns are numbered 1 to 6. Column 2 could be recorded in such a way as to differentiate, if necessary, between 'men', 'women' and 'children'. Further differentia-

| Date | Number of Men | Operation | Crop/Livestock | Hours of Work | Remarks |
|---|---|---|---|---|---|
| 1 | 2 | 3 | 4 | 5 | 6 |

Figure 9.1 *Labour Use Book.*

tion may specify between 'hired', 'communal' and 'family' labour, especially if wages have to be paid for hired labour and particularly so if there is no separate Wages Book.

In column 4 it is possible to have the operation for a field in a rotation containing several crops (e.g. yams and maize). In that case both crops have to be entered. And when the work is maintenance (e.g. 'cleaning store house'), then the column 4 entry should read 'store' or 'machine' or, more specifically, 'tractor' as the case may be.

*Farm Purchases and Sales Book*

An example of a page in this type of record book is given in Figure 9.2. The columns are numbered 1 to 6. Where there are many transactions per day either way or where seasonality of purchases is different from

| PURCHASES | | | SALES | | |
|---|---|---|---|---|---|
| Date | Particulars | Value | Date | Particulars | Value |
| 1 | 2 | 3 | 4 | 5 | 6 |

Figure 9.2 *Farm Purchases and Sales Book.*

the seasonality of sales, it may be desirable to open two separate books: one exclusively for purchases and another for sales. Values in columns 3 and 6 have to be in money terms, except where gifts that may have to be recorded are involved, in which case such transactions will be recorded in another book (either Crop Yields and Disposals Book or the Livestock Record Book or even the Livestock Products and Disposals Book).

*Crop Yield and Disposal Book*

A specimen page of this type of record book is supplied in Figure 9.3. The relevant columns are numbered 1 to 8. Column 3 may be used for

| | CROPS RECEIPT | | | | CROPS DISPOSAL | | |
|---|---|---|---|---|---|---|---|
| Date | Crop | Average (from whom) | Total Yield | Date | Crop | Quantity | How/To Whom Disposed |
| 1 | 2 | 3 | 4 | 5 | 6 | 7 | 8 |

Figure 9.3 *Crop Yield and Disposal Book.*

'from whom' especially in cases when the crop is either 'purchased' or a 'gift'. Similarly, column 8 may be used to indicate 'how' or 'to whom' the crop is 'sold', given as 'gift' or otherwise used (e.g. as 'livestock feed'). Column 4 may sometimes refer to quantity instead of yield, especially for crops not cultivated on the farm.

*Livestock Record Book*

This record book will only be necessary and useful on a farm where livestock enterprise is an active part of the farming operation. An illustration of the information that such a record book may contain is given in Figure 9.4, with the columns numbered 1 to 11. Column 10, which indicates the total change for any specified data, will have to be with a positive (+) sign for net addition to stock and a negative (–) sign if otherwise. Entries may be daily, weekly or even monthly, depending on convenience and the number of recordings to be done.

In making entries in column 2, 'adult male' may be separated from 'adult female' and 'young' from 'adult' for each row of entries made. In other words, it is better to have different rows for the different age group and sex of each kind of stock kept.

*Livestock Products and Disposals Book*

A specimen page ruling for this record book is illustrated in Figure 9.5, with columns numbered 1 to 8. Columns 3 and 7 will be for either quantity (e.g. of milk) or number (e.g. of eggs).

| Date | Kind of Stock | Previous Number of Stock on Hand | Additions (+) | | | Losses (−) | | | Net Change (+) | Final Stock Position |
|---|---|---|---|---|---|---|---|---|---|---|
| | | | Bought | Born | Gifts | Sold | Death | Lost | | |
| 1 | 2 | 3 | 4 | 5 | 6 | 7 | 8 | 9 | 10 | 11 |

Figure 9.4 *Livestock Record Book (for sheep/cattle/poultry/others).*

| INCOMING | | | | OUTGOING | | | |
|---|---|---|---|---|---|---|---|
| Date | Product Kind | Quantity or Number | Source | Date | Product Kind | Quantity or Number | How Disposed |
| 1 | 2 | 3 | 4 | 5 | 6 | 7 | 8 |

Figure 9.5 *Livestock Products and Disposals Book.*

*Cash/Bank Account Book*
Where a bank account is kept with a significant number of transactions, it may be better to have a separate Cash Book and Bank Account Book. The ruling of such books will be as illustrated in Figure 9.6, numbered with columns 1 to 5. The ruling in the figure is simple but functional. It can be made to show more details, such as price per unit of sales or purchase or even quantities/numbers, as the case may be, for future reference. All these can be included under 'particulars' in column 2 or 5. Extra columns may also be added if it is considered necessary.

| RECEIPTS | | | EXPENDITURES | | |
|---|---|---|---|---|---|
| Date | Particulars | Amount | Date | Particulars | Amount |
| 1 | 2 | 3 | 4 | 5 | 6 |
| | | | | | |

Figure 9.6 *Cash/Bank Account Book.*

If the Bank Account Book is to be part of the Cash Book, then all deposits made to the bank have to be entered in the expenditure side of the book as 'to bank' (amount). Similarly, all withdrawals from the bank have to be entered in the receipts side of this book as 'from bank' (amount). The difference between the left hand totals and the right hand totals in this book will be a cash balance. This balance is positive if receipts are greater than expenditure, and negative if otherwise. Such a negative cash balance is possible if a large amount from non-farm sources has been put away in the bank and entered in this book. To avoid many embarrassing mistakes it is advisable to record every financial transaction immediately and in the appropriate column.

## 9.3 CLOSING AND BALANCING BOOKS

Finally, at the end of every convenient period, all these books have to be 'closed'; that is, the totals have to be made and checked. When this is being done it is useful to 'take stock' of all enterprises. It is useful to check that book balances agree with physical balances, and mistakes or disagreements between the two have to be reconciled. Once the books are checked and reconciled, then the final asset of the farm activities for the period (quarter, half-year or year) can be financially estimated or ascertained. To be able to do this, a special account called the Profit and Loss Account has to be opened. Where the farming enterprise is elaborate and highly commercialised, it may be necessary also to prepare a Balance Sheet, which is to give the assets (i.e. the worth of the belongings) of the business for a specified period, usually a year. The Balance Sheet is extracted from the Profit and Loss Account, and it is usually considered to be the job of professional book-keepers or accountants.

### *Profit and Loss Account Book*

This account book is usually opened at the beginning of the farm year or cropping season, whichever is more convenient, and closed at the end of the period. It is a very important account, since it summarises all assets, liabilities, expenses and revenue for the stated period and thus gives a clear situation of the profitability or otherwise of the farming activities within the specified period.

On a relatively small farm, with one or two lines of interest, all activities for the year can be summarised in one account book. On the other hand, a farm with a number of enterprises and interests may call for different Profit and Loss Account Book for the separate enterprises as well as for a Balance Sheet to give a summary of all interests in an aggregate manner.

The essential information usually included in a simple Profit and Loss Account Book is given in Figure 9.7.

The Profit and Loss Account Book has two sides. The left hand side (LHS) is concerned with liabilities and payments. These include the values of all items of goods, properties and materials available and used for the farming activities for the specified period. The right hand side (RHS), on the other hand, includes the values of all items left at the end (or closing) of the same period.

### *Stock Valuation*

The question of 'opening' and 'closing' valuations raises a problem. This is concerned with whether, as a result of use or of other farm activities,

| Liabilities and Payments | | Assets and Revenue | |
|---|---|---|---|
| Items | Value | Items | Value |
| (1) Opening valuation<br>– Livestock<br>– Crops (store & field)<br>– Other physical assets (buildings, dams, equipment, machines, plants, etc.) | | (1) Closing valuation<br>– Livestock<br>– Crops (stored & field) | |
| (2) Other investment | | (2) Present value of other investment | |
| (3) Labour (wages) | | (3) Sales–crops–livestock | |
| (4) Other input costs (fertiliser, seeds, chemicals, feedstuffs, fuel, etc.) | | (4) Other sales | |
| (5) Repairs & maintenance | | (5) Salvage (spare parts) sales (value) | |
| (6) Insurance & taxes | | (6) Rebate or discount | |
| (7) Debts to pay (e.g. interest on loans) | | (7) Outstanding income revenue | |
| (8) Other overheads (electricity, water, telephone, security, etc.) | | (8) Interest on other investment | |
| (9) Opening cash balance | | (9) Closing cash balance | |
| (10) Opening bank balance | | (10) Closing bank balance | |
| (11) Balance brought down (B/D) (if there is *profit*) | | (11) Balance carried forward (C/F) (if there is a *loss*) | |
| (12) Total (LHS) | | (12) Total (RHS) | |

Figure 9.7 *Profit and Loss Account Book for the year ending . . . (or for the period . . . to . . .).*

there has been an improvement in the market value of these items of stock or depreciation (or even obsolescence, i.e. not of any further use) of the stock in part or as a whole. Where the value of the goods has reduced or *depreciated* in value, the closing valuation of that stock has to be lower by the estimated amount of depreciation. Machines, plants (not crop plants) and equipment usually depreciate with use. The value of depreciation, for the sake of simplicity, is calculated by dividing the purchase price by the length (number of years) of possible use:

$$\text{value of depreciation} = \frac{\text{value at purchase}}{\text{no. of years of use}}$$

On the other hand, where the stock is *appreciating*, e.g. a growing animal or permanent crop that is yet to attain maturity, the closing valuation can be calculated by either of two methods, and it is usually advisable to use the lower figure so as to avoid inflating the profit or assets of the farm. This is important too because of taxation on profits.

One method of calculating 'closing valuation' for growing stock is to use the estimated market value. The other method is to use the 'opening valuation' of the stock and add the 'value of additional expenses' since that last valuation. Such additional expenses may include 'feeding stuffs and labour costs' in the case of livestock, or 'labour costs' in the case of growing crops.

*Balancing the Books*

The other important issue to be emphasised has to do with the 'balance' (item 11) on both sides of the account page. After items 1–10 on both sides have been entered, a totalling up is done. If the total on the right hand side is greater, this reflects that the assets of the farm exceed the liabilities for the specified period; on that ground a gross profit is indicated. If, on the other hand, it is the total on the left hand side that is greater, then a loss is indicated. Therefore, whichever side is greater, the difference has to be entered on the side with the smaller total as 'balance' on only one side of item 11. After this is done the final totals (item 12) on both sides have to be equal. When this procedure is completed the book is said to be *balanced.*

*Gross and Net Profit*

*Gross profit* is defined in a simple way as the difference between the assets and liabilities of the farm within a specified period. But *net profit* for a farm where the farmer is the manager will be the balance after the management cost has been deducted from the gross profit. In practice it is always difficult to separate the day-to-day existence and activities of the farmer on his farm and other lines of activities. Hence, for all practical purposes the profitability or otherwise of the farm is usually assessed on a gross profit basis.

*Chapter 10*

# Producing and Selling Farm Products Profitably

## 10.1 DEMAND, SUPPLY AND PROFIT

The science of *economics*, which is concerned with the use of scarce resources for the satisfaction of wants, is of great significance in all commercial activities undertaken for profit. One particular economic aspect of relevance to profitable farming is the law of *demand* and *supply*. This law in its simplest form refers to three aspects of the production (supply) and sales (demand) of all goods and services:

(1) When more goods or services are available (supplied) than are wanted or can be sold (demanded) the *price drops*.
(2) When more goods or services are wanted (demanded) than are available (supplied), the *price rises*.
(3) When the quantity of goods or services available (supplied) is the same as the quantity wanted (demanded), then the *price remains constant*.

In farming, these three aspects of the law of demand and supply will apply to the purchase of inputs (or *factors of production*) as well as to the farm products available for sale. Therefore, for profitable farming it is always essential to make sure that *revenue* exceeds the *cost of production*, leaving a *profit*:

$$\text{revenue} = \text{quantity of product sold} \times \text{price}$$
$$\text{cost of production} = \text{quantity of input used} \times \text{price}$$
$$\text{profit} = \text{revenue} - \text{cost of production}$$

## 10.2 PROFITABLE PRODUCTION

One of the major economic principles useful in achieving profitable production is to produce the right goods (i.e. the goods demanded) as cheaply as possible; this principle is sometimes called the *cost minimisation* principle. The other economic principle of importance, related to profitable sales, is to sell as much as possible at as high a price as possible to achieve maximum revenue; this principle is sometimes referred to as *profit maximisation* (or maximum profit possible).

In farming for profit, therefore, the two important points to bear in mind are: (a) to try to produce as much as the market wants at as low a cost as possible; and (b) to try to sell as much as possible (market demand) for as high a price as possible.

For any commercial farmer, i.e. a farmer producing profitably for the market (for sale), there are three stages to follow: (a) the thought process; (b) decisions to be made; and (c) actions to be taken.

### 10.21 THE THOUGHT PROCESS

Farming is a difficult occupation. It calls for hard work, most of which is influenced or dictated by weather, climate and the environment. Therefore going into farming needs careful thought; making a profit of farming needs more careful thought; and being successful at farming calls for still more careful thought.

In thinking about farming and making a profit from farming, or better still being a successful farmer, it is important: (a) to collect as much information as necessary; (b) to analyse and try to understand as much as possible from the information collected; and finally (c) to keep a good record of all useful information.

The necessary information to seek, and sometimes pay for, can be grouped into four: (a) the *weather and ecology* of the area to be farmed (e.g. rainfall, temperature); (b) the *technical facts* about farming and the crops or livestock to be handled (e.g. soil, planting dates, varieties/breed, crop yields, milk yields, egg yields, weight of food consumed, fertiliser to use); (c) the *type and cost of inputs* or resources to be used (e.g. land rent, labour wages, fertilisers, seeds, machinery) and the range (over the year) of the *market selling price* for the anticipated farm products; and finally (d) the *government legislation, organisations and assistance* offered for the type of farming activity to be undertaken.

### 10.22 DECISIONS TO BE MADE

Taking decisions is also a very difficult task, especially when money and perhaps livelihood are involved. In farming, whether for profit or for home consumption, a number of important sets of decisions have to be taken. There are at least eight important sets of decisions to be made in profitable farming, as follows:

#### *What Enterprise to Undertake*

The decision on what enterprise (or line of activity) to follow or undertake will be influenced by a few important factors, some of which are: (a) ability and interest of the farmer; (b) consumers' interest; (c) soil of the area to be used; (d) ecology and what it can be encouraged to produce; (e) availability of inputs for the type of production (e.g. land); and (f) availability of market for the products (including by-products).

*Where to Operate*

The decision on where to operate for profitable location will also be influenced by the need to locate the enterprise with respect to its distance from the city on a rational idea, as indicated in the zonal diagram in Figure 10.1.

From the diagram it can be seen that products that have high market value and are perishable are located near the city, while heavier commodities (e.g. tubers and root crops) or products that lose weight after processing are situated farther away from the town. The appro-

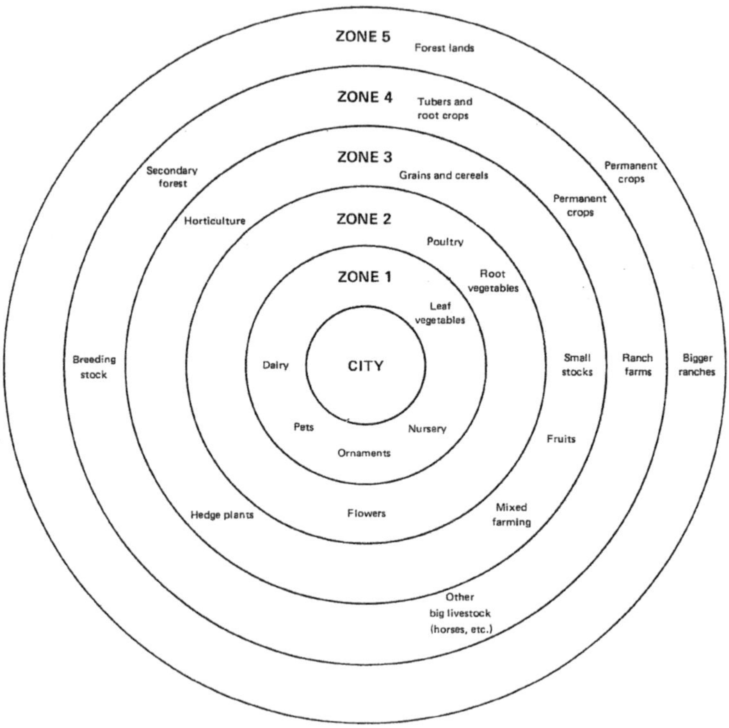


Figure 10.1 *Zones of farm activities.*

priate distance for each zone will depend on the size of the town, cost of purchasing land, degree of economic activity, and other factors which we do not need to go into here.

*How Much to Produce*

Decisions on how much to produce will depend on: (a) available suitable land; with (b) corresponding available other inputs (e.g. labour, fertiliser) and costs; and (c) market (quantity and price).

*Method of Production*

Decisions on the profitable method of production will depend on: (a) type of enterprise being undertaken; (b) available capital (cash and equipment); (c) size of farm (stock size); (d) technology available; and again (e) market.

*Farm Operations*

Decisions on farm operations are complicated but have to be planned well in advance. Some of the important decisions on farm operations include the following:

(1) *Procurement of inputs to the farm:* source, amount needed, when, how, costs.
(2) *Preplanting operations* (e.g. bush clearing, ridging): their timing, how to be done, by whom, inclusive of nursery operations.
(3) *Planting operations:* their timing, seeds required (quantity, type, where available, cost), time and method of transplanting.
(4) *Cultural (post-planting) operations* (e.g. thinning, supplying, weeding, grafting, budding, topping tobacco).
(5) *Pre-harvesting operations* (e.g. barn erection or repair, store houses, cleaning, weighing equipment, harvesting baskets, labour).
(6) *Harvesting operations:* when, how, labour needed.
(7) *Post-harvesting operations* (e.g. handling of produce, sorting, bagging, packaging, storage, processing, selling).
(8) *Selling:* best market, timing, how to get produce to market, grades.

*Market Operations*

Decisions on market operations will take into consideration: (a) quantity of product available for market (marketable surplus); (b) which market will give the best possible price and when; (c) when to go to market; (d) costs of marketing (e.g. transport, agents' commission) and (e) anticipated revenue and profit level.

*Storage and Processing*

Decisions on storage and processing are closely related. That part of the farm output not to be consumed and not immediately available for the market has to be stored or processed. Hence decisions on storage include the following:

(1) *How much to store*, depending on the facilities available for storage, the cost of storage, the change in price over time, and the profits to be gained by storing.
(2) *For how long to store*, depending again on the cost of storage and the profit to be gained by storing.
(3) *Method of storage*, depending on the available know-how (technology), available facilities and profit.

Similarly, decisions on processing include the following:

(1) *How much* is available for processing.
(2) *How much* is needed to be processed.
(3) *When* the processing is to be done.
(4) *By whom, where* and at *what cost* (after examining the alternative processing methods and their respective costs).
(5) *What extra profit* is realised after processing:
extra profit=(total revenue after processing)−(cost of product now+cost of processing+profit now).
(6) *Delivery to market:* packaging, transport, labour.
(7) *Selling*: price, markets, and in what units to sell for best price and maximum profit.

*Raising Livestock*

Most of the decisions mentioned above will to a large extent apply also to livestock and husbandry men. However, there are a few important decisions specifically related to breeding that have not been mentioned. These include decision with respect to:

(1) *Extent of livestock activity:* whether it will concentrate on raising livestock products (e.g. eggs, meat, milk), or on raising breeding stock, or on raising breeding stock as well as livestock products. It is essential to focus attention on the cost and returns of attractive breeding systems or of raising livestock products such as milk, meat, and eggs.
(2) *Breed to use*, depending on the desirable qualities of the male and female animals (and/or birds) relevant to the products the farm will be selling.
(3) *Size of herd or poultry stock*, depending on available funds, market, type of product to sell, etc.

## 10.3 ESSENCE OF SUCCESSFUL (EFFICIENT) FARMING

There are three major aspects of successful farming: (a) making the right decisions, (b) timely action, and (c) good management.

*Making the Right Decisions*
To make the right decisions, the farmer has to explore all available information on the technical as well as the economic aspects of the farm operations. Once this information is available, then taking the right decisions can be simplified. The sets of decisions discussed in section 10.2 by no means cover all the relevant issues. However, if most of these are thoroughly analysed, the job of successful farming is well started.

*Timely Action*
Decisions should lead to action. In farming, more than in any other occupation, taking action at the appropriate time is highly essential for any profitable return. Therefore relevant decisions should be made well in advance of the anticipated time of action. All actions should be carefully and thoroughly executed to fit in with the necessary ecological, biological and economic factors that influence all farm activities and operations.

*Good Management*
Successful farming means profitable farming with satisfaction for a job well done. For any profitable business, farming included, there is no substitute for good management. That is, on the one hand, all available resources should be used adequately, timely and efficiently to produce the best products and as much as possible. In addition, farm products should be disposed of at the appropriate time for as much as possible with a minimum of waste or delay.

Management is very important in four main areas: (a) in buying things for the farm; (b) in using resources (e.g. labour, land, machines, animals) to produce things on the farm for sale; (c) in selling products for the farm; and (d) in keeping good, accurate and regular records of all important farm activities and operations. If all these are done well, to the best of the farmer's ability, then success is a matter of time.

*Part Six*

# THE IMPROVEMENT OF WEST AFRICAN AGRICULTURE

*Chapter 11*

# Agricultural Improvement at the National Level

## 11.1 IMPORTANCE OF AGRICULTURE

If we define the discipline of *agriculture* in its broadest sense to include the sub-fields of farming, pastoral activities (including nomadism), forestry and fisheries, then a study of the measures relating to agricultural improvement will have to cover all these fields.

The relative importance of agriculture to a country, especially the developing countries – West African countries – in particular, includes the following:

(1) Agriculture provides food, clothing (e.g. cotton, wool) and shelter (e.g. timber, palm fronds).
(2) Agriculture provides employment both (a) directly, as farmers, cattle rearers, hunters, sawyers of forest timbers, and fishermen; as well as (b) directly, in producing cutlasses, hoes, guns or baskets for use on farms and in the processing, storage and distribution (marketing) of agricultural products.
(3) Some farm products (e.g. rubber, palm oil, palm kernels, cocoa, tobacco, cotton, milk) are processed in local industries, thus providing raw materials for industries.
(4) Industries provide fertilisers, equipment, chemicals and machines for use on the farms too. Thus agriculture directly provides market (demand) for industrial products in addition to the demand of farmers and other agriculturists for processed products and consumer goods. This helps to boost local industries and indirectly helps the pace of industrialisation of the country.
(5) Surplus labour (population) on the farm tends to move into suburban and urban areas and provides relatively cheap labour to work in factories and other sectors of the economy of the country.
(6) Agriculture has an important role to play in the export trade of any developing country, supplying cocoa, coffee, groundnuts, rubber, oil seeds, timber, hides and skins to other countries – mostly to the developed world, where these products are processed into manufactured products.
(7) Most of these exports constitute the main source of foreign exchange (other currency) for the country. This foreign exchange is useful

in paying for the country's imports from other countries – most significantly for the import of machinery, equipment and skilled workers, all of which are necessary for the country's economic improvement and development.

(8) As part of other advantages of agriculture to the country, mention must be made of the training on the farm as helping to develop management ability in other occupations; in providing sporting interests (hunting); in helping education by preserving wildlife in the forests; and in attracting tourists to see natural wildlife and game.

(9) Agriculture, including forestry, if properly managed helps in preserving the natural ecological balance between all living things, especially between man and his environment. For instance, forests help to improve rainfall, check erosion and increase soil micro-organisms.

In view of all these benefits that any country gets from correct and well-planned agricultural practices and activities, it is very important indeed that the development of agriculture should be given great priority.

There are two sets of activities by government and other organisations in any country that affect the progress, improvement and development of agriculture: (a) specific national efforts for agricultural improvement; and (b) other national efforts that indirectly have important effect on agriculture.

## 11.2 SPECIFIC NATIONAL EFFORTS FOR IMPROVING AGRICULTURE

*Provision of Improved Farm Inputs*

Materials for planting (e.g. improved seeds, fertiliser, chemicals) or for breeding have to be improved and made cheaply and readily available for all farmers in the country. To do this, scientists have to be encouraged and instructed by government to produce better, high-yielding, climatically adaptable and disease-resistant types and varieties of seeds and livestock for use in the country.

Farming makes use of tools, equipments and machines. These have to be developed and be made adaptable to local uses and skills. Scientists, technicians and engineers in the country have to be encouraged to design these things for the improvement of farming, and factories have to be set up to produce these things for use on farms.

*Training In Improved Farming Methods*

There are different types of farming and a number of farm activities.

If these activities are carried out more efficiently, there is bound to be an increase in farm output, and hence more food, more livestock products and more export crops will be available. These increased products will mean that the citizens of the country will eat better, be healthier and still have more money from exports.

The easiest way of bringing these improvements about is to study how the farmers at present carry out their activities and from this be able to appreciate their problems. Then the government would be in a position to find ways and means of improving their farming methods and their standard of living, by training these farmers in better ways of farming.

### *Imparting Skill*

*Skill* can be defined simply as the ability to do something or carry out an activity. On most farms in West Africa, as in most developing countries, poor skill makes it difficult for the farmers to do better than they do at present. Most of them do not know how to store their products for the market. Only a few of them can use simple machines or complicated farm equipment, which if properly used can help to improve their farming. Therefore every government has to make an effort to teach skill in handling machines and farm products and in carrying out good farm operations to all farmers.

In an effort to do this, the government must set up institutions, training schools and agricultural colleges where farmers and agriculturists can be trained and given the necessary skills for improving their farm operations.

### *Agricultural Extension*

*Agricultural extension* is the out-of-school teaching of farmers in ways and means of doing a better job of farming. There are also *agricultural extension services*, which include the services providing farmers with materials and equipment they may require to assist them in better farming. Both definitions can be used as synonyms.

The governments in most developing countries spend a lot of money and use a number of trained personnel for their extension services. This is done to bring to farmers modern knowledge, techniques, and improved seeds and materials which are to help in improving farming and farmers' way of life all over the country. Unfortunately, because of the few and poorly trained workers used by governments for the extension services, most farmers do not benefit as much as they should from these extension services. Governments in their efforts to improve agriculture through their extension *agents* (as the workers are called) have to train their agents very well so that they can help farmers in a more practical way to solve the major farming problems that farmers have to battle with from day to day. Governments, by providing adequate improved seeds and materials, equipment and machines as well as funds and financial encouragement (inducements) for agricultural extension workers, can bring about improved farming and a better way of life for all working in and concerned with agriculture in their countries.

*Agricultural Research*

If agriculture is to improve, it is necessary to have knowledge of better methods of farming and improved seeds, breeds, materials, equipment and machinery to use. All these call for intensification of agricultural research. Research will advise the governments, and subsequently through extension services the farmers, about alternative methods and farm inputs that are available. Research will also bring out the relative costs and returns of these alternatives. The more and the better the alternatives made available to the farmer, at least up to a point, the better the agriculture and the easier it is to improve and develop the agriculture of the country.

Improving agricultural research in any country is a necessary first step to agricultural development. Government must provide funds, well-trained and well-paid research personnel, up-to-date research equipment, laboratories and research institutions such as Agricultural Universities, as well as build up an efficient means or network of transmitting research findings to farmers and their fields. All these things will cost money and have to be carefully planned and well executed. Without them agriculture is not going to develop, and farmers will continue to be poor, starving and unproductive.

*Agricultural Institutions and Organisations*

These have to be developed along two main lines. They have to be improved (a) to cope with new situations and new lines of farm activities, and (b) to carry out the job they are expected to do more efficiently.

In every society there are institutions, for instance, for having access to farmland, labour, farm credits and such facilities. With changes in the society and human groups and ways of life, some of these (traditional) institutions become either inadequate or not able to handle things in a modern fashion. They have to be modified, improved or replaced by new institutions that can operate better.

Similarly, farmers have their traditional organisations or societies and their own accepted ways of doing things. However, over time there are changes both in the farmers' relationships and groupings that sometimes call for changes in the type and scope of these organisations. Governments must be quick to notice these trends and help in bringing about the changes in the farm development projects, settlement schemes, etc. that are necessary and beneficial.

*Markets (Exports and Domestic) for Agricultural Products*
Improving the marketing systems and institutions will have four advantages:

(1) It will help to encourage farmers to produce more if they realise that they can sell all they can produce at reasonably good prices in the market.
(2) It will enable people in urban areas to buy enough quantities of farm products both cheaply and at the time they need them. At the same time, manufacturers of farm products will be in a position also to buy the raw materials needed for their factories at the desirable price, quantity and quality.
(3) It will increase the amount of exports, and consequently the farm income and foreign exchange, through the increased trade that results from improved marketing and a developed agriculture.
(4) The farm inputs (seeds, fertilisers, etc.) and other farm purchases will, through improved marketing, become cheaper and readily available to the farmer.

*Industry for Processing Agricultural Products*
This type of industry is called *agro-based* or *agro-allied* industry. The establishment of industries for the manufacturing of agricultural products will, if well planned and organised, have four benefits for the community and especially for the farmers:

(1) Processing factories will give to the farmers an assurance that, provided they cultivate what the factory wants, their products will be purchased and they will have some money to spend.
(2) Processing factories will provide some employment, which will help people in the rural areas if such factories are based in rural areas.
(3) Processing factories indirectly help to improve the standard and organisation of farms, because the factories will lay down directions and instructions on how best to produce what the factory can buy and use.
(4) Processing factories will save for the country all the foreign exchange the people would have spent in buying the new processed goods from abroad; thus the country will be in a position to buy other materials and equipment needed for the development of its economy.

*Provision of Credit*
Every business needs money (sometimes called *capital*), which can be in the form of *cash* or *credit*, to buy necessary goods and services for

the use of the business. In agriculture the problem is a serious one, for three reasons:

(1) Farmers are very poor people and do not have the means of saving enough capital to put at the use (or disposal) of their farming business. Most of the time, farmers in West Africa have barely enough money to meet their day-to-day expenditure on the basic necessities of life (e.g. food, clothing, adequate shelter).
(2) The alternative source of capital is by borrowing money from a bank or other commercial lending house. This is not readily available to farmers, either (a) because there are no such facilities nearby; or (b) because, when available, the bank requires a guarantee of repayment (a *bond* or *collateral* or *surety*) before the loan is given; or (c) simply because the bank does not approve the loan, regarding it as unprofitable (i.e. the bank regards the use for which the loan is required as not meriting attention from their commercial point of view).
(3) Farmers are forced to borrow money from money lenders at exorbitant rates of interest, with the *mortgage* or the transfer to the money lender of the farmer's land, property, future harvests, or even his own or his relations' labour until the debt is paid. In this way the farmer is perpetually in debt. As a result of such debt the farmer is made poorer and poorer, and his life and living conditions become steadily worse.

For these three main reasons at least, and others not discussed here, it is vital for the development of agriculture in any country that the government makes adequate arrangement to guarantee that farmers have capital at a cheap rate of interest available to them. This capital need not all be in the form of cash. Some can be in the form of equipment (e.g. hoes, ploughs, spray pumps) or in the form of inputs (e.g. seeds, fertilisers, livestock). Sometimes credit can be made available in the form of *subsidies* on these necessary farm materials or on equipment that the farmers need.

Different forms of organisations such as Thrift and Credit Co-operatives, Co-operative Banks, Agricultural Credit Banks and Farmers' Credit Associations are being set up in different countries, to help in ensuring that capital credit is made available to farmers readily and cheaply and that the credits that are granted are used mainly for productive purposes. This type of arrangement is referred to as *supervised credit.*

### *Agricultural Festivals and Shows*

In many countries, farmers are traditionally relaxed people because they live so close to nature. Nevertheless they always like to get together, essentially to exchange information about all aspects of farming and also to use the opportunity for a social get-together during which some food, amusements and competitions are introduced for additional excitement.

Farm shows, agricultural festivals, cinema shows, farming on television and radio farm programmes are ways and means of bringing farmers together and of passing farming information to a large group of them for their benefit. It is therefore important that governments should not only provide enough funds for these types of extension activities but also accept them as an active part of farm life, and pursue the policy forthrightly.

In some cases it is desirable to appeal to the younger generation by introducing Young Farmers' Clubs and gardening in schools and Agricultural Hobby Groups in urban areas, and not wait until farming becomes too unproductive before making frantic efforts to encourage people to farm.

Governments must plan ahead, provide adequate funds and personnel, encourage participation, and allow farming to have a pride of place in the social, economic and political life of all age groups in both the urban and rural areas of the country. It is only by so doing that governments can encourage farming and make it profitable. When all these things are done, then agriculture will be in a position effectively to provide all (food, employment, clothing, shelter, etc.) that is expected of it.

## 11.3 OTHER NATIONAL EFFORTS THAT INDIRECTLY IMPROVE AGRICULTURE

Under this heading can be grouped all government efforts that improve the physical, social, political and economic well-being of all citizens. These are of course as important as those discussed in section 11.2, but they only affect agricultural improvement in an indirect way. Some of these efforts include the following:

### *Planning the Economy*

One of the most difficult tasks, and the most vital, in the development of any economy is how to plan such development. In planning any economy there is need to know (a) what resources of all types (human, land, materials, etc.) the country has, and (b) how best to combine them to increase the production of what the citizens need or can sell to others. In so doing, agricultural production will also have to be planned, and the planning of agriculture is only meaningful if it leads to improved farming and an increased income and standard of living for farm people.

*Building of Farm Roads*

Without enough roads leading from the city to the villages, there will be poor communication. On the other hand, improved farm roads will make it easier, cheaper and timely to get supplies to the farms. At the same time, such roads make it possible to get farm products quickly to the cities and other places (including other countries) where these products can be used. Besides, people will prefer to live on farms if they have the ease of being able to get to the cities and other places when and as often as they want.

*Provision of Water, Electricity, Hospitals, Schools, etc.*

These amenities are called *infrastructure.* If they are provided in villages and to farm people generally, they will indirectly improve agriculture in the following manner:

(1) Farm people will live more healthy lives; hence they will be able to work harder and produce more and better farm products to eat and sell.
(2) More and more farm people will become educated, learn about science, acquire skills, learn how to do things better and be able to lead a more satisfactory life and be better and happier citizens. This will indirectly also improve their attitude to farming and thus help them to produce more.
(3) It will also reduce the urge or attraction to go to cities to live. Many farmers and their children leave the farm for the cities, going because of the lack of amenities in the rural areas, and the few that do not move are not pleased to stay on; hence they do not produce of their best.

*Employment in Rural Areas*

There are several ways of increasing the opportunity for other jobs and occupation in rural areas. When there are other jobs besides farming, people without farmland of their own or those who do not like farming can be employed and will earn some money. Then they will be more prepared to stay in rural areas instead of travelling to the cities to look for jobs. If such rural employment is well planned, it can indirectly encourage the improvement of agricultural production.

*Chapter 12*

# Agricultural Improvement at the Farm Level

In Chapter 11 the areas of government efforts to improve agriculture at the national level were carefully examined. In this chapter we want to concentrate on how agricultural improvement can be achieved or induced at the farmer's field or at the farm level. The efforts have to be controlled partly by the government, but mainly so by the farmer himself – including his family of course. Hence emphasis in this chapter will be on the contribution of the farm family towards a rapid improvement of agriculture in developing countries generally and West Africa in particular.

In general, the necessary improvements at the farm gates are influenced by two main areas of knowledge: (a) the effect of science and technology on improved agriculture; and (b) the impact of economics and business methods on farming.

## 12.1 EFFECTS OF SCIENCE AND TECHNOLOGY ON FARMING

There are five main areas in which the application of science and technology helps the farmer to improve his farm operations: (a) with inputs; (b) better techniques of farm operations; (c) farm buildings and machines; (d) ecological balance; and (e) health, nutrition and the environment.

### *With Inputs*

Improved seeds; better breeds of livestock for higher yields or greater resistance to disease or drought; fertilisers for improving soil fertility as well as increasing yield of crops; chemicals for the control of insect pests, internal and external pests of animals, weeds, etc. are all of great importance and made available to agriculture through the tireless efforts of scientists.

### *Better Techniques of Farm Operations*

Improvement can be brought about by way of better planting dates, adequate spacing, optimum plant population, soil analysis for a thorough understanding of the potential as well as the deficiencies of farmland, and also in other farm operations (e.g. mulching, budding, grafting). Methods of soil cultivation, handling of farm animals, etc. are important in that if properly carried out they can mean the difference between success and failure. Science and scientific methods continually look into ways and means of improving these farm activities.

*Farm Buildings and Machines*
Through the help of engineering (i.e. the science and technology of engines), different kinds and forms of farm buildings and structures as well as tools, equipment and machines are made available to help the farmer with his numerous farm tasks. Whether it is a storage bin, or a dam for irrigation, or a tractor for farm cultivation, there is always a wide range of choice to suit the different needs and money at the disposal of the farmer. Continuous efforts are being made to improve these machines, etc.

*Ecological Balance*
Ecological balance is very important, with respect to water for instance. Where there is not enough water, the farmer can provide irrigation by using any one of the different methods available. By contrast, where the soil is waterlogged, several drainage methods are known and equally available by which the farmer can improve his farmland and increase his yields, as well as bring more waterlogged areas under cultivation for increased output.

In another sense, the balance between crop land, forest, livestock, human needs and the human population can be adequately adjusted. In most countries in Africa, shifting cultivation and the bush fallow system are well-known methods for restoring soil fertility. They are also useful for creating ecological balance. However, the problems of population and shortage of farmland in some parts of West Africa have jointly made it difficult to keep on with these practices, and science has come in to help with this problem by providing fertiliser and organic manures.

*Health, Nutrition and the Environment*
The prevalence of diseases, problems of poor nutrition and unhealthy environment give rise to low farm output. It is important for the family to be in good health to be able to engage actively in farming. The farmer, his family and livestock have to eat well, i.e. adequate amounts of all required nutrients regularly, to have enough energy for the farm work, to build enough body tissues and to be able to produce more. The importance of a clean healthy environment is often forgotten. It has to be remembered that disease-causing organisms and pests thrive better in dirty places.

## 12.2 ECONOMICS AND BUSINESS METHODS IN FARMING

The second area in which farming can be improved is through the efforts of the farmer in applying economics and business principles, methods and organisation in his farming activities. For successful farming, skill is necessary. But in addition it is essential to have business management knowledge and experience. The application of both skill and efficient management will surely guarantee profitable and successful farming. However, there are many aspects to the farming business.

### 12.21 ECONOMIC DECISIONS

Most farmers in West Africa started farming initially for subsistence. Now that the structure of the economy is changing, there is a growing need to produce more and more with the market in view. Hence the economic decisions necessary to improve the performance of the farmer in response to the modern needs for all kinds of agricultural products are urgent and vital. Economic decisions will include decisions on what to produce, where, when and how much to produce. In strict economic terms, these decisions will lead to action that will call on the use of resources to produce the farm products. In so doing, money will be spent. Such expenditures will only be considered good business ideas or propositions if the return from the sales of the products far exceeds the money spent. In other words, profit is made in that the gross revenue exceeds total expenditure. If the farm business is not profitable, the future of the business becomes uncertain, and continued losses because of bad decisions or poor judgement will eventually lead to a collapse (closure or bankruptcy) of the business.

### 12.22 BUSINESS APPROACH TO FARMING

One of the most important, and perhaps the weakest, areas in traditional farming in West Africa is the lack of business skill, organisation and management. Without the right business attitude to farming the farmer cannot hope to make a profitable success of farming.

Four important business approaches to farming are: (a) securing business information; (b) the ability to weigh and take necessary and calculated risks; (c) efficient organisation and good management; and (d) effective sales promotion.

*Securing Business Information*

In farming, as in any other business, it is wise to study the economic prospects, methods, procedures, code of conduct and other important areas related to how to become a successful businessman in the area of operation selected. Getting some of this information may cost money. Nevertheless, the alternative of the present system of guesswork, hit and miss is too expensive and only practicable and rational for very small businesses.

*Risks and Uncertainty*

Perhaps the most crucial element in business is taking risks and being able to gauge how much risk to take. To be good at taking risk needs some experience. It is also helpful to be able to talk to those who are in the business for guidance, or even to read about what to do and what to avoid doing. Finally, it is important to weigh the advantages and disadvantages of the steps being taken, i.e. to weigh the *degree of risk* involved.

In farming there is the additional problem of uncertainty about the weather, the condition and quantity of farm products to expect, and the market situation for such products. Again, the only effective way of being able to overcome or reduce the effect of uncertainty in farming is by experience by, talking to other farmers and knowledgeable officials of the Ministry of Agriculture, and by reading scientific materials on the relevant line of business. Sometimes, of course, governments help to reduce uncertainties in farming through price guarantees and crop insurance schemes.

*Organisation and Management*

Improvements of the farm business depends to a large extent on the efficiency of organisation and management of the farm and farm operations. The organisation of the farm will include both (a) the *physical location* of farm buildings, pens, farmland, pastures, etc.; and (b) *operations* such as timing, labour requirements, tools, materials and equipment. The management aspect, on the other hand, is concerned with making the best possible use of all the available resources and products of the farm. Efficient use and well-organised farm operation will help to reduce the per unit cost of producing farm products (cost minimisation), thereby indirectly raising the level of profit.

*Effective Sales Promotion*

This may sound most unusual in agriculture, even though every business organisation knows the value and importance of sales. Sometimes it is more important to be able to sell more at a lower profit per unit. This helps to make a full use of resources and to expand or increase the size of the business.

In agriculture, as in any other business proposition, increasing the volume of sales is important. This can be achieved through: (a) advertisement by newspaper, radio, television, signboards, etc.; (b) distribution of samples; (c) personal contacts to people likely to want to buy (i.e. prospective customers); (d) producing specially good quality, graded, standardised products; (e) good personal attention to *all* customers; and (f) good marketing organisation and management of products, agents and customers.

All said and done, the few issues discussed so far have been limited to the more essential areas. It is important to stress that for effective agricultural improvement the national efforts have to be simultaneous with efforts at the farm level. They are inter-related and should be attacked together and effectively.

# Glossary

*abattoir:* Slaughter house where animals are killed and cut into parts.

*abdomen:* The part of the body posterior to the thorax (and oesophagus) containing the digestive organs.

*abomasum:* The fourth chamber of the stomach of ruminants; others are the rumen, reticulum and omasum.

*abortion:* Premature birth taking place before the end of the full pregnancy period.

*adaptation:* The process by which an organism becomes adjusted to an environment.

*albumen:* The clear part of an egg surrounding the yolk; white when boiled or fried.

*alimentary canal:* see *digestive tract.*

*amino acid:* Organic acid containing the amino (nitrogen) group $NH_2$; amino acids are constituents of all proteins.

*androecium:* The male reproductive organs of a seed plant.

*annual plant:* A plant that completes its lifecycle within one calendar year.

*anther:* The part of a stamen that produces pollen in a flower.

*antibiotic:* A substance produced by micro-organisms that prevents the growth of other organisms or bodies.

*antibody:* A protein produced in an organism to counteract the action of a foreign substance.

*antigen:* A foreign protein that induces the formation of an antibody in an organism.

*artificial insemination:* The process by which sperms that have been collected artificially from a male animal are injected through a tube into the vagina of a female on heat.

*ascorbic acid:* The pure water-soluble vitamin C.

*ascus* (plural *asci*)*:* A spore sac characteristic of the Ascomycetes fungi, usually containing eight spores called *ascospores.*

*assimilation:* The transformation of food substances into protoplasm.

*bacterium* (plural *bacteria*)*:* A microscopic single-celled organism not visible to the naked eye.

*beef:* Meat of cattle.

*beriberi:* A disease due to lack of vitamin $B_1$.

*bladderworm:* A stage in the development of a tapeworm.

*broody hen:* A hen wanting to sit on eggs.

*bud:* An undeveloped shoot.

*bulb:* A short underground stem surrounded by fleshly leaves that store food.

*bulbil:* An aerial bud or bulb that may detach itself from parent material to produce a new plant.

*caecum:* A blind pouch from some part of the digestive tract.

*calciferol:* Another name for vitamin $D_2$.

*calfing:* The process of giving birth in cattle.
*calyx:* The sepals of a flower.
*carbohydrate:* An organic compound consisting of carbon, hydrogen and oxygen.
*carotene:* A red orange pigment found in the plastids of some plants.
*carpel:* A flower's female organ, made up of a stigma, style and ovary.
*carrying capacity:* Ability of a pasture to feed adequately during grazing a given number of farm animals within a specified time.
*castration:* The removal of the testes from a male animal to prevent him from reproducing.
*cell:* The fundamental unit of all living things; composed of protoplasm which consists of a nucleus, cytoplasm and a bounding plasma membrane; surrounded in plants by a cell wall.
*cellulose:* A complex carbohydrate.
*cell wall:* A rigid cellulose structure surrounding and regulating the shape of plant cells.
*cercaria:* A trematode larva with a tail (e.g. tapeworm).
*cervix:* The narrow structure of an organ (e.g. uterine cervix).
*chalaza:* (a) Bands attaching the yolk to the shell membrane of a bird's egg; (b) a structure of an ovule.
*chlorophyll:* Green pigment found in the chloroplasts of plants; important in photosynthesis.
*chloroplast:* A plastid containing chlorophyll, found in the cytoplasm of cells in the green parts of the plants.
*chromosome:* A thread-like body in a cell nucleus containing genes.
*clay:* Soil type with particules less than 0·002 mm in diameter.
*cloaca:* A common chamber into which the digestive, excretory and reproductive systems open in birds.
*coleoptile:* A sheath that encloses the plumule of the seedlings of monocotyledons (e.g. maize, wheat).
*colon:* Part of the large intestine in animals.
*colostrum:* Milk produced immediately after parturition; contains high levels of protein, minerals, vitamins and antibodies.
*corm:* A short enlarged underground stem that stores food (e.g. cocoyam).
*corolla:* The petals of a flower.
*cotyledon:* A seed leaf.
*crop:* (a) A thin-walled enlarged part of a bird's digestive system immediately after the oesophagus; (b) agricultural plants on a farm.
*cross-breeding:* The breeding of two different breeds to produce hybrid offspring with characters of both parents; used in animal improvement programmes, especially for the hybrid vigour; contrast to in-breeding.
*culling:* Picking out and killing poor and weak animals.
*curing:* Treating, for example, tobacco by drying (*flue-curing* or *fire-curing*) or meat by salting or smoking.
*cytoplasm:* Protoplasm outside the nucleus of a cell.

*dicotyledon:* A plant having two cotyledons.
*digestion:* The conversion of complex food substances into simpler compounds which can be readily absorbed.

*digestive tract:* The passage from mouth to anus wherein digestion occurs; the alimentary canal.
*diploid:* The double set of chromosomes usually found in somatic cells; twice the haploid number.
*duodenum:* The first part of the small intestine, where bile from the gall bladder and pancreatic juices mix with the food.

*ecology:* The study of organisms in relation to their environment.
*embryo:* A young organism (animal or plant) in the early stages of development.
*endocarp:* The innermost layer of a fruit wall.
*endosperm:* A nutritive tissue outside the embryo of some flowering plants.
*epidermis:* The outermost layer of an organism.
*erosion:* The wearing away of rock particles or soil, by the action of water, wind, rain or temperature.
*excretion:* The removal of poisonous or waste products from the body.
*exocarp:* The outermost layer of a fruit wall.
*exoskeleton:* A hard supporting structure external to and covering parts of the body (e.g. in insects).

*fallow:* Land that is not being used for cropping.
*farm yard manure (FYM):* Type of animal residue comprising of dung, urine and bedding that have been allowed to rot together to produce a useful manure for use on farms.
*farrow:* The process of giving birth in pigs.
*fecundity:* Ability to bring forth offspring regularly.
*fermentation:* A process by which energy is released through the action of certain organisms in the absence of oxygen.
*fertilisation:* The union of the nuclei of male and female gametes to form a zygote.
*fertiliser:* A substance of known chemical content, which is added to the soil in order to improve it.
*fertility:* Ability to produce male or female sex cells regularly, leading to successful mating in a regular manner.
*filament:* (a) The stalk of a stamen of a flower; (b) a slender thread.
*foetus:* A young mammal within the uterus of the mother.
*free range system:* A method of keeping livestock whereby the animals roam all over the place without restraint.
*fungicide:* A chemical substance used for destroying fungus diseases.
*fungus* (plural *fungi*)*:* A plant composed of a mass of thread-like forms, which lacks chlorophyll and is therefore unable to make its own food.
*furrow:* A long cut between two ridges made by a plough.

*gall bladder:* A gland in the liver which produces *bile* for digestion in the duodenum.
*gamete:* A reproductive cell which can unite with another reproductive cell of a different sex to form a zygote.
*gene:* A hereditary unit of living things; contained in a chromosome.

*geotropism:* A growth movement in plants induced by gravity.
*germ spot:* The disc-like area of an egg yolk in which the first segmentation takes place if the egg is fertilised.
*gestation:* The period between fertilisation and giving birth; *pregnancy* period in mammals.
*ginning:* The removal of the lint from cotton seed by hand or machine.
*gizzard:* A muscular chamber in the alimentary canal of birds, useful in physical grinding of solids.
*glume:* The outer part of a grass spikelet.
*graft:* A part of a living tree which is inserted into another tree.
*gynoecium:* The female reproductive organs of a seed plant.

*haploid:* The single set of chromosomes usually found in gametes; half the diploid number.
*haustorium:* A hypha, usually with fungi, which functions as a penetrating and absorbing organ.
*heat period:* A period when a female is ready and shows signs for mating.
*herbicide:* A chemical substance used for destroying weeds on farms.
*heredity:* The passing of characters from parents to offspring through sexual or asexual reproduction.
*hermaphrodite:* An organism that possesses both male and female reproductive organs.
*heterosis:* The result of a cross between (two different zygotes) heterozygous parents; usually done to increase vigour and improve the breed.
*heterozygote:* An individual with different genes in a gene pair.
*homozygote:* An individual with similar genes in a gene pair.
*host:* An organism that provides nourishment or protection for another organism, usually called a parasite.
*humus:* Decomposed organic matter in the soil; rich in nitrogen.
*hybrid:* The offspring of plants or animals differing in one or more genes.
*hypha* (plural *hyphae*)*:* One of the filaments of the mycelium of a fungus.
*hypocotyl:* The part of a seedling below the cotyledons.

*immunity:* Ability to withstand disease attack which is acquired either from having suffered from the disease or from artificial treatment.
*inbreeding:* The breeding of animals in successive stages such that the offspring have more and more of a particular gene (increasing homozygosity); contrast to cross-breeding.
*ingestion:* The taking in (eating) of food.
*insecticide:* A chemical substance used for destroying insects.
*irrigation:* Supplying the soil with water for the use of plants.

*kiln:* A furnace for drying or baking purposes, usually at high temperatures.
*kraaling:* The art of keeping farm animals on harvested plots on the farm; the animals feed on the remains of the harvest and in return leave their dung and urine on the farm to be ploughed into the soil for the following crop.

*lactation:* The secretion of milk from the udders of female animals, or its extraction by hand or milking machine.

*larva* (plural *larvae*): An early stage after the egg in some animals' development.
*ley:* A grass field suitable for grazing or used as cover.
*loam:* A soil composed of sand, silt and clay in fixed balanced proportions.
*lodging:* Toppling over of trees or plants.

*macronutrient:* An element that is required in large quantities by plants or animals; opposite of micronutrient.
*management:* The science and art of looking after things on the farm with a view to obtaining better results from them.
*manure:* Natural or artificial food for plants, added to the soil to make plants grow better.
*meiosis:* The reduction division of cells which halves the chromosome number from diploid to haploid; occurs in the production of gametes; see also *mitosis*.
*mesocarp:* The middle layer of the fruit wall.
*metabolism:* The chemical process by which food material is built up or broken down in the cells of an organism.
*metamorphosis:* Major structural changes by which certain animals develop from an embryo to an adult (e.g. insects).
*micronutrient:* An element that is required in small quantities by plants or animals; opposite of macronutrient.
*mitosis:* Cell division involving the duplication of chromosomes and the formation of daughter nuclei with the same number of chromosomes as the original cell nucleus; see also *meiosis*.
*monocotyledon:* A plant having one cotyledon.
*mulch:* A cover of grass, leaves etc. spread out around plants to retain moisture, smother weeds and protect plants from the direct heat of the sun.
*mutton:* Meat of sheep.
*mycelium:* The mass of hyphae comprising a fungus.

*nitrification:* The oxidation of nitrogen compounds, ammonia and ammonium salts into nitrites and nitrates by certain bacteria.
*nitrogen fixation:* The conversion of free nitrogen of the air into complex forms.
*node:* A point on a plant stem at which leaves and buds may arise.
*nucleus* (plural *nuclei*): A cell structure containing the chromosomes and surrounded by a *nuclear membrane* and cytoplasm.
*nutrition:* The ingestion, digestion and assimilation of food.
*nymph:* An early stage in the development of an insect which differs from the adult only in size and in the absence of wings.

*oesophagus:* The tube leading from mouth to stomach.
*ovary:* (a) The organ that produces ova in animals; (b) in plants, the ovule-containing part of a carpel in a flower.
*ovule:* A female structure containing a gamete which grows into a seed after fertilisation.
*ovum* (plural *ova*): Female gametes or eggs.

*pancreas:* A gland which produces digestive enzymes and *insulin* for digestion in the duodenum.

*panicle:* A branched inflorescence of some complexity.

*parasite:* An organism that lives on or within the body of another living organism from which it obtains part or all of its food; *ectoparasites* on the outside; *endoparasites* within.

*parturition:* The delivery of offspring at the end of pregnancy.

*pasteurisation:* A process by which food (e.g. milk) is heated to a temperature of about 70 °C (158 °F), which will kill most active bacteria.

*pathogen:* Any disease-producing micro-organism or substance.

*perennial plant:* A plant that lives for more than one year.

*pericarp:* The outside wall of a fruit.

*pest:* An organism that is harmful to man, his crops or his animals.

*petal:* One of the leaves (often coloured) of the corolla of a flower.

*phloem:* The food-conducting tissue in a vascular plant, consisting mainly of *sieve tubes.*

*photosynthesis:* The process by which green plants utilise the energy of sunlight, bringing together carbon dioxide and water to form carbohydrates.

*phototropism:* A growth movement in plants induced by light.

*pith:* The tissue in the central part of a stem.

*placenta:* A structure in the uterus linking maternal and foetal tissues, serving for the respiration and nutrition of the growing young.

*plasma membrane:* The outermost part of protoplasm; forms the boundary in animal cells.

*plastid:* A small dense protoplasmic inclusion in a cell; often pigmented.

*plumule:* The first bud appearing during germination.

*pollen:* Powder produced by the anthers of a flower; contains male gametes.

*pollination:* The transfer of pollen from a stamen to a stigma.

*protein:* A nitrogen-based organic compound composed of amino acids and containing carbon, hydrogen, oxygen, sulphur and phosphorus.

*proven stock:* A good animal that has proved its capability to transfer some of its good characteristics to offspring in a reliable manner.

*proventriculus:* A chamber in the alimentary canal of birds.

*protoplasm:* The living liquid substance of plant and animal cells.

*pupa* (plural *pupae*)*:* A resting stage in the metamorphosis of an insect, after the larva. The pupa develops into an adult.

*quadruped:* An animal with four legs that are all used for moving about.

*quarantine:* The keeping of plants and animals that have recently been introduced apart from the original group until there is no risk of the newcomers spreading disease.

*radicle:* The part of a plant embryo that develops into the primary root.

*ration:* The quantity of food fed to an animal in a specified time period, usually a day: (a) *balanced ration* contains all the necessary elements in their right proportions; (b) *maintenance ration* is just enough food to maintain only the body activities of the animal; (c) *production ration* is the additional ration given to the animal for producing some products (e.g. milk, eggs, work); or even for mating; (d) *steaming-up ration* is given to stimulate good development of the foetus inside the mother as well as to increase milk supply at delivery.

*receptacle:* The portion of a plant to which the flower parts (male and/or female) are attached.
*reproduction:* The process by which any organism produces an offspring.
*respiration:* The process by which an organism obtains energy through the oxidation of organic substances within it.
*rhizome:* An underground horizontal stem which is useful in propagating the plant.
*roguing:* Uprooting a plant entirely to prevent it spreading disease.
*rotation of crops:* Different crops are grown one after the other in a sequential order on a piece of land.
*ruminants:* Animals that virtually survive mainly on green plants (grasses, etc.) because they have the digestive system to handle the fibres in such green plants.
*runner:* A stem which grows horizontally along the ground but not underneath the soil.

*sand:* Soil type with particles between 0·02 and 2 mm in diameter.
*saprophyte:* An organism that obtains its food mainly from dead organic matter.
*savanna:* A vegetation type characterised by natural grasslands with a few scattered trees.
*scolex:* The head of a tapeworm parasite.
*scutellum:* The single cotyledon of a grass (embryo) seed or fruit.
*sepal:* A leaf-like part of the outermost structure of a flower.
*sheath:* see *coleoptile.*
*shell membrane:* An organic layer containing the albumen and yolk of an egg, inside the shell.
*silage:* The green herbage or cut plant tissues that are stored in a *silo* to preserve the green colour until they are needed for feeding animals.
*silt:* Soil type with particles between 0·002 and 0·02 mm in diameter.
*somatic cell:* A non-reproductive cell of the parent body.
*sperm:* The male gamete which is capable of moving to fertilise the ovum.
*spike:* An inflorescence with an elongated main axis (*rachis*) and flowers (*spikelets*) round it.
*spore:* A non-sexual reproductive cell, usually protected against adverse conditions.
*sporocyst:* An encysted embryo stage of trematodes (e.g. tapeworm).
*stamen:* A flower's male organ, made up of an anther and a filament.
*stigma:* The pollen-receiving part of a carpel in a flower.
*stock:* (a) *Livestock* is all animals kept on a farm; (b) *deadstock* is equipment, machines and non-living equipment; (c) *cropstock* is plants growing and plant harvests stored on a farm.
*stoma* (plural *stomata*)*:* A small opening in the epidermis of a leaf.
*style:* The slender part or channel of a carpel of a flower, supporting the stigma.
*subsistence farming:* The production of crops, forest trees and livestock solely for home consumption.
*sucker:* A stem-branch developed below ground and ultimately forming a new plant (e.g. banana).
*symbiosis:* An association of two different kinds of living organisms which is of benefit to both.

*tanning:* Changing of hides and skins into soft leather by treatment with tannic acid, washing and drying.
*testa:* The outer coat of a seed, usually of cellulose.
*testis* (plural *testes*): The male reproductive organ which produces sperms; in mammals usually two.
*thorax:* The part of the body between the neck and the abdomen, called the *chest.*
*tillering:* Production of side shoots by a plant; usual in Monocots (grass Family).
*toxin:* A poisonous substance derived from a plant or animal.
*trace element:* Equivalent to micronutrient; an element required in small amount.
*traction:* Act of pulling or driving a load or machine.
*transhumance:* The movement of man and his belongings, including cattle, from place to place primarily for survival.
*transpiration:* The evaporation of water from leaves or other wet plant surfaces.
*transplanting:* The transfer of young plants from a seed bed or nursery to the final bed where they will develop fully.
*tuber:* An enlarged short underground stem or root used for storing starches.

*udder:* The milk-producing (*mammary*) organ of certain female animals.
*umbilical stalk:* The cord which connects the placenta to the foetus in the uterus.
*upgrading:* The improvement of farm animals by mating them to better breeds.
*uterus:* The enlarged part of the female reproductive system in which the foetus develops.

*vaccine:* An attenuated living culture of pathogens which is injected into the body to protect it against a disease.
*vagina:* The terminal part of the female reproductive system, leading from the uterus to the genital opening.
*vascular bundle:* A tissue containing xylem and phloem in roots, stems and leaves.
*vector:* An organism that carries and transfers pathogenic organisms.
*virus:* A sub-microscopic particle which can be cultivated only in living cells; composed of a nucleic acid core and a protein covering.
*vitamins:* Organic substances which are necessary in small amounts for the normal growth and function of plants and animals; deficiency is known as *vitaminosis.*

*waterlogged soil:* Soil that is completely full of water; not well drained.
*whey:* The liquid part of milk after the separation of curds; useful in feeding pigs and adult calves.

*xylem:* A vascular tissue containing vessels that are adapted for conducting water from the roots to the leaves.

*yolk:* The nutrient reserve of an egg.

*zygote:* A cell resulting from the fusion of two gametes of different sexes; grows into an embryo.

# Index

For Product Safety Concerns and Information please contact our EU
representative GPSR@taylorandfrancis.com
Taylor & Francis Verlag GmbH, Kaufingerstraße 24, 80331 München, Germany

www.ingramcontent.com/pod-product-compliance
Ingram Content Group UK Ltd.
Pitfield, Milton Keynes, MK11 3LW, UK
UKHW022155190926
475702UK00003B/45
* 9 7 8 0 3 6 7 2 5 0 0 2 7 *